PRAISE FOR **Subtract**

"This is a captivating, perceptive read on one of the most basic mistakes that we make in the pursuit of success and happiness. Combining his expertise in architectural engineering and behavioral science, Leidy Klotz pinpoints a gaping hole in our mental math: we're constantly adding tasks, commitments, and possessions to our lives while neglecting to subtract any. If the defining word of your life is 'more,' you need to read this book."

—Adam Grant, *New York Times* bestselling author of *Think Again* and *Originals*, and host of the TED podcast *WorkLife*

"The heart of Leidy Klotz's *Subtract* is a brilliant catch-all philosophy for living well: humans are predisposed to add—money, ideas, inventions, possessions, complexity—but the route to well-being is to take away. Subtraction doesn't require you to be rich or to have superhuman self-control or plenty of free time. All you need is the right attitude, and a toolbox that will help you learn to subtract effectively. *Subtract* is that toolbox."

—Adam Alter, *New York Times* bestselling author of *Irresistible* and *Drunk Tank Pink*

"A good book whisks you on a journey to foreign lands filled with exotic ideas and captivating stories. A great book changes the world you live in, pulling back the curtain to reveal mysteries you didn't even know were there. This is a great book.

Anyone who has interest in understanding their lives better—and who doesn't—should read it."

—Sendhil Mullainathan, MacArthur fellow and coauthor of *Scarcity*

"Leidy Klotz leads the reader on an intellectual journey as he explores the fascinating question: In designing our world, why do we always seem to add rather than subtract? *Subtract* is a great blend of academic research, stories, and practical tools. Enjoy the journey!"

—Dan Heath, coauthor of *Made to Stick* and author of *Upstream*

"Behavioral science at its best helps people to improve their lives. It may help them think about old things in new ways or add new and better habits. In *Subtract*, Leidy Klotz shows us how deleting things from our lives can lead us to exciting new places."

—Carol Dweck, author of *Mindset*

"*Subtract* is simply brilliant. With engaging and moving prose, Leidy Klotz travels back and forth between kids' toys, evolution, Dr. Seuss, anthropology, racism, cognitive psychology, urban planning, global warming, and architecture to teach us that, often, the best way to solve problems is by taking things away—by subtracting. To learn how to subtract, you must first add *Subtract* to your reading list."

—Barry Schwartz, author of *The Paradox of Choice* and *Why We Work*

"There could be no more important time to read this well-researched book, when so much needs subtracting: the prejudices, norms, and rules that perpetuate structural racism in our country; the mindless consumption that puts us on the hedonic treadmill and destroys biodiversity and a livable climate on our planet." —Elke Weber, Gerhard R. Andlinger Professor in Energy and the Environment, Princeton University

"An engaging style and breadth of examples that guide the reader through an important new view of thinking by removing."
—Eric J. Johnson, author of *The Elements of Choice*

Subtract

Subtract

The Untapped Science of Less

Leidy Klotz

FLATIRON BOOKS
NEW YORK

Printed in the United States of America. For information, address Flatiron Books, 120 Broadway, New York, NY 10271.

www.flatironbooks.com

Designed by Donna Sinisgalli Noetzel

The Library of Congress has cataloged the hardcover edition as follows:

Names: Klotz, Leidy, 1978– author.
Title: Subtract : the untapped science of less / Leidy Klotz.
Description: First edition. | New York, NY : Flatiron Books, 2021. | Includes bibliographical references.
Identifiers: LCCN 2020053766 | ISBN 9781250249869 (hardcover) | ISBN 9781250249937 (ebook)
Subjects: LCSH: Self-actualization (Psychology) | Stress management.
Classification: LCC BF637.S4 K5557 2021 | DDC 158.1—dc23
LC record available at https://lccn.loc.gov/2020053766

ISBN 978-1-250-24987-6 (trade paperback)

First Flatiron Books Paperback Edition: 2022

10 9 8 7 6

For Josephine

Contents

Introduction

The Other Kind of Change

1.

When my family visited San Francisco, the Embarcadero waterfront was at the top of our to-do list. We saw the historic piers and the Ferry Building. We strolled along a palm tree–lined promenade. We looked for wild parrots in a lush park. A balloon artist made a monkey for my son, Ezra, who toddled safely to show it to the harbor seals. Lots of other people seemed to be doing things they'd always remember.

It had taken an earthquake to make this unforgettable place. Well, an earthquake with some help from a woman named Sue.

Before the Embarcadero was a must-see destination, it was a double-decker concrete highway. Like so many other city-crossing highways in the United States, the Embarcadero Freeway was built after World War II, made possible by federal support for highways to move the military and serve the growing number of automobiles.

For decades after it was built, the Embarcadero Freeway had stretched more than a mile along San Francisco's eastern waterfront, blocking precious views and access to the bay. Elsewhere in the city, community organizers—initially downplayed as "little housewives" by pro-highway groups—had stopped plans for highways that would have done more harm than good to the city. But the freeway along the Embarcadero was serving tens of thousands of vehicles per day. It was one thing to determine that a new highway was unnecessary; it was another to ask whether it might be a good idea to remove a freeway that had already been built. Fortunately, San Francisco had Sue Bierman.

After growing up in Nebraska, Sue Bierman had moved to San Francisco in the 1950s, bringing with her more formal training in music than in city planning. But Bierman was incredibly bright and driven, a high school valedictorian and incessant reader. She had learned how to get things done in San Francisco as a community organizing housewife. Based on her success in that role, she earned an official appointment, in 1976, to the city's planning commission.

Bierman was meticulous in her public service. Her planning commission studied the Embarcadero Freeway using all sorts of metrics: how much traffic it carried, how many customers it brought to city businesses, how the freeway affected property values, and how it affected quality of life—in the neighborhoods it connected and in the neighborhoods it crossed. The commission also considered options for how they might transform the existing freeway. What were the costs and benefits of turning the double-decker freeway into a subsurface tunnel? Of extending the freeway so that it connected to the Golden Gate bridge? Of leaving well enough alone and focusing on other parts of the city? It took all that analysis and more, over nearly a decade, for Bierman's commission to finally, in 1985, offer their plan for the Embarcadero Freeway: get rid of it.

Businesses near the freeway opposed the plan, worrying that reduced automobile traffic would mean fewer customers. More surprisingly, in retrospect at least, businesses weren't alone in resisting. When San Franciscans voted on the proposal to remove the freeway, it wasn't even close. For every voter in favor of removing the freeway, there were two who wanted to keep it. Whether for fear of traffic, fear of lost business, or simply fear of change, voters rejected removal. The people had spoken. Sue Bierman and her commission moved on to other projects.

The Embarcadero Freeway might still be blocking San Francisco's waterfront if not for the Loma Prieta earthquake, which struck on October 17, 1989. As a sports-obsessed middle-schooler tuning in expecting to see the third game of baseball's World Series, I experienced the earthquake as millions of others did, first a blank screen, then panicked announcers broadcasting from San Francisco's Candlestick Park, and then collapsed highways and a burning city, all on live television.

The Loma Prieta earthquake killed more than sixty people and injured thousands. A concrete slab as big as a basketball court fell from the upper to the lower deck of the Bay Bridge. Fires swept through the Marina District, just a few blocks north of the Embarcadero. People sat stranded outside next to whatever they could gather before evacuating their homes. The earthquake caused about $6 billion in property damage alone. At the time, it was the most expensive earthquake in the history of the United States.

The earthquake changed the calculus for removing the Embarcadero Freeway. First, the post-earthquake freeway had been rendered unusable. To repair the damaged and aging structure, to have it carry traffic again, was going to cost far more than knocking it down. And second, the earthquake was a tragic warning of the risks of elevated freeways. Many of the people who had died in the Loma Prieta earthquake had been crushed when the Cypress Street Viaduct collapsed in Oakland. And as a double-decker elevated concrete structure just over a mile in length, the Cypress Street Viaduct looked ominously like the Embarcadero Freeway.

Still, even in this post-earthquake reality, plenty of smart San Franciscans wanted to revitalize their ruin. Engineers suggested it be repaired, reinforced with thicker concrete columns, and otherwise remain in place. Local businesses agreed, and so did many residents. The Pulitzer Prize–winning *San Francisco Chronicle* columnist Herb Caen, for whom the walking promenade along the freeway-free waterfront is now named, opined, "Once again, there is 'serious talk' about tearing down the Embarcadero Freeway, an even worse idea than building it."

This time, however, there would be no popular vote, which may have kept the freeway in place. Instead, the decision was put to the city's board of supervisors, who, by the narrowest possible six-to-five margin, finally approved the planning commission's original recommendation.

Sue Bierman didn't have long to gloat. In 1991, she was relieved of her duties by the new mayor, who was keeping his winning campaign promise to get rid of the planning commissioners that had gotten rid of the Embarcadero Freeway.

It had taken an earthquake and some sacrificial public servants, but the freeway came down. With it removed, tourists and San Franciscans got the waterfront back. The decade after removal saw a 50 percent increase in housing and a 15 percent increase in jobs around the waterfront, far outpacing gains in other parts of the city. Demolishing the freeway did not cause traffic nightmares, as some had predicted. Trips were rerouted, to the grid of surface streets, to other access ramps to the Bay Bridge, and to public transit. People found new ways to move about the city. The corridor that used to cater exclusively to automobiles now serves as many walkers as it does riders.

For those who have visited, such evidence is redundant. It is obvious why the Embarcadero should not be covered with a freeway. By 2000, the ten-year anniversary of the freeway's demolition, the *San Francisco Chronicle* was reporting that it had become "hard to find anyone who thinks ripping down the Freeway was a bad idea."

Upon her death, a newspaper eulogy hailed Sue Bierman

as "the quintessential neighborhood activist" across her half-century of service to San Francisco. Today, Sue Bierman Park, near the waterfront and surrounded by the hectic financial district, is a five-acre green oasis where my son, Ezra, and I looked for wild parrots.

Around the time Sue Bierman was making a final push for the Embarcadero Freeway to come down, dockworker Leo Robinson was being thanked by Nelson Mandela during a speech to around sixty thousand people packed into the Oakland Coliseum. Robinson was an unlikely dismantler of oppression in South Africa.

Born in Shreveport, Louisiana, in 1937, Robinson and his family moved to the Bay Area when he was a young boy, drawn by the promise of better opportunities for African Americans than in the Deep South. Yet the Robinsons lived in a de facto segregated, "redlined," neighborhood and Leo's parents found jobs only because of a perfect storm of executive orders from the president, organized protests against discriminatory hiring, and a labor shortage during the booming wartime economy. Robinson dropped out of high school in twelfth grade and joined the navy, serving in the years after the Korean War.

After an honorable discharge, Robinson went to work on the docks in the early 1960s. Initially, he didn't concern himself much with affairs beyond his own, let alone those in South Africa. When tracing the genesis of his political activism,

Robinson recalled a conversation when he felt unqualified to offer an opinion on the United States' role in the Vietnam War. Robinson immersed himself in politics and it wasn't long before he was taking action on a variety of global issues.

One of Robinson's targets became the apartheid system in South Africa. Apartheid's racist policies evoked his own struggles with segregated neighborhoods, discriminatory hiring, and income inequality in the United States. Emphasizing these parallels, Robinson helped establish and grow an anti-apartheid group of dockworkers.

When the ship *Nedlloyd Kimberley* docked at San Francisco's Pier 80 late in 1984, Robinson and his group of dockworkers unloaded most of the ship and then moved on, refusing to unload the South African cargo. The "bloody" steel, auto parts, and wine—all of it sat on board, untouched. Robinson's group was so powerful that none of the nearby ports would accept the apartheid cargo either.

As Robinson had hoped, the dockworkers' refusal set off a series of anti-apartheid actions. Thousands of people took part in daily protests next to the stranded *Nedlloyd Kimberley.* Soon, the city of Oakland had pulled all of its funds out of companies that did business in South Africa. The state of California followed Oakland's lead, reallocating more than $11 billion that had been invested in South Africa. Similar divestment spread to other cities, states, and nations. Multinational companies including General Electric, General Motors, and Coca-Cola rushed to sever ties with apartheid South Africa.

There had long been organized resistance to apartheid, especially within South Africa; that was why Nelson Mandela and so many like him had been in jail. But once international divestment started, apartheid's days were numbered. That's why, when Mandela spoke in Oakland, he thanked Leo Robinson and his fellow dockworkers for being the "front line of the anti-apartheid movement in the Bay Area."

Leo Robinson improved a social system. Sue Bierman improved a city. And Elinor Ostrom improved an idea. Ostrom's gift was more in carving away from the idea than it was in extending it—as is often the case with Nobel Prize winners such as herself.

Over a career devoted to studying economic governance, the Indiana University professor Ostrom whittled away at a theory known as "the tragedy of the commons." The tragedy theory had been championed by the ecologist Garrett Hardin, who in an influential 1968 essay, had revisited an old fable about herders with grazing cattle on common land. Each herder has to decide how many cattle they will allow to graze on the land. If every herder restrains themselves to a reasonable number of cattle, then the common land will have time to replenish itself each year, and therefore support generations of herders. But there is a dilemma: If some herders limit their cattle but others do not, then the common land is exhausted, and the herders who limited themselves have sacrificed the short-term benefits of taking a bigger share. It follows that, even if you wish to be an unselfish herder, if you know that other herders are selfish, then you might as well take as much as possible, and fast. Get it while you can.

Hardin extended the herding analogy to modern environmental issues. Selfish herding behavior would prevail, he figured, whenever there are resources useful to a lot of people but not owned by any of them. Many environmental issues can be viewed as common dilemmas—including climate change, with our life-supporting atmosphere as the common resource, and humans burning fossil fuels into deadly amounts of greenhouse gases as the selfish herders. Hardin argued that the only way to deal with this frequent scenario, the way to stave off environmental devastation, was through private ownership of natural resources.

Hardin's tragedy builds from assumptions about human motivation, about the rules governing the common-pool resource, and about the resource itself. Ostrom showed that these assumptions are wrong. Humans are quite capable of managing the commons without tragedy. With her meticulous field research, Ostrom discovered that it happens all over the world: in Indonesian forests, Nepalese irrigation systems, and New England lobster fisheries.

Whereas Hardin proposed a general theory from a fable, Ostrom distilled more nuanced themes from the evidence. One is that tragedy can be averted with a combination of community care for the resource (as in lobsterers who self-police overfishing in conversations at the local bars), and larger-scale governance (as in federal laws that threaten to stop all lobstering if the species becomes endangered).

Elinor Ostrom's gift to collective knowledge was an edit. She started with Hardin's proposal that common-pool situations are destined for tragedy—and she showed that each unique situation is more like a drama. With thoughtful planning, Ostrom found, we can write our own happier endings.

What these three have in common is that they tapped into the power of *subtraction*. Sue Bierman subtracted a freeway to create one of the most visited places in the world. Leo Robinson sparked the financial subtraction that brought down apartheid. Elinor Ostrom subtracted wrong ideas to give humanity a better approach to our common future. All three made positive change because of their thought, courage, and persistence in taking away. And all three made change because they saw opportunities everyone else had missed.

2.

Do your resolutions more often start with "I should do more of . . ." than with "I should do less of . . ."?

Do you have more stuff than you used to?

Do you spend more time acquiring information—whether through podcasts, websites, or conversation—than you spend distilling what you already know?

Do you spend more time writing new content than editing what's there?

Have you started more organizations, initiatives, and activities than you have phased out?

Do you add new rules in your household or workplace more often than you take rules away?

Do you think more about providing for the disadvantaged than about removing unearned privilege?

Are you busier today than you were three years ago?

If you answered yes to any of these questions, you're not alone. In our striving to improve our lives, our work, and our society, we overwhelmingly *add*. As we'll see in the pages to come, there are many interwoven reasons for this—cultural, economic, historical, and even biological. As we'll also see, it doesn't have to be this way.

To be sure, sometimes more is better. When my family returned from our trip to San Francisco, we spread out in a home with a new five-room addition. In other cases, adding makes sense in isolation but crowds us over time, such as when the first floor of our home addition filled with tens of thousands of Legos. And sometimes taking away brings delight. No longer is my running confined to a treadmill, from which I used to listen to books and podcasts while watching the news on television, leaving my brain with no chance to turn data to knowledge to wisdom. The rewards of less came only after evidence honed my thinking about how to get there.

Whether it's Sue Bierman surveying San Francisco's waterfront, me considering my home renovation, or you making resolutions, we're all doing essentially the same thing—trying to change things from how they are to how we want them to be. And in this ubiquitous act of change, one option is always to add to what exists, be it objects, ideas, or social systems. Another option is to subtract from what is already there.

The problem is that we neglect subtraction. Compared to changes that add, those that subtract are harder to think of (quite literally, as we'll see in the next chapter). Even when we do manage to think of it, subtracting can be harder to implement. But we have a choice. We don't have to let this oversight go on taking its toll on our cities, our institutions, and our minds. And, make no mistake, overlooking an entire category of change takes a toll.

Neglecting subtraction is harmful in our households, which now commonly contain more than a quarter of a million items. Someone has to organize and keep track of all those juicers, ill-fitting clothes, Legos, and long-since-deflated monkey balloons from family trips to San Francisco. That's a lot to pay for and to think about, and it represents a lot of our time, time that is only getting scarcer, especially when we overlook subtraction as a way to relieve our obviously overbooked schedules.

We neglect subtraction in our institutions. In our governments and in our families, we default to adding requirements. Ezra gets more and more rules, and grown-ups deal with federal regulations that are twenty times as long as they were in 1950. Too many rules and too much red tape can distract from the behaviors we're really hoping for, in our kids and, as we'll see, in our dairy farmers. What's more, when it comes to social change, the subtractive options we are missing are often the better ones. Donating money to anti-apartheid rebels is helpful, but it doesn't remove the harmful system's power. Taking money away from apartheid does.

When it comes to how we make meaning of the world, subtraction neglect is so strong that experts describe learning as "knowledge *construction*." When we have a misconception in our knowledge, constructing more on top of it is the mental equivalent of trying to reinforce an earthquake-damaged freeway. Ideally, we would get rid of the outdated idea and build on stable ground. Yet, as individuals and as a society, having learned Garrett Hardin's tragedy we tend not to question it—which is especially harmful in this case because Hardin was a eugenicist who used his tragedy theory to argue against a multiethnic society. Regardless of where ideas originate, to keep our minds open, we need to do the counterintuitive work of removal.

Neglecting subtraction is even bad for the planet. As Dr. Seuss recognized nearly half a century ago in his environmentalist classic *The Lorax,* if we hope to leave options, we need to subtract stuff. Now, when there is more carbon dioxide in the atmosphere than is safe, we can't just add it more slowly (though that would be a good start). We also need to subtract some.

There is good news for *Subtract* readers. By understanding the nature and roots of our adding, we can learn how to find less across disparate worlds. And if you can be among the few who subtracts, you will exploit the inefficiency in our market of changes.

3.

This book is the manifestation of my longstanding obsession with less. At least since I was a teenager, I've pondered what seemed to me like a widespread inability to get rid of things that weren't making anyone's life any better. Back then, my summer job was to mow grass, a task that gave me lots of time to think, often about the need for lawns that only seemed to be used when I was mowing them.

Two decades removed from my lawn boy career, I think about subtraction while playing with my son, Ezra. Does he have more books and more Legos and more distractions than I did at his age? Why, since long before he could string together coherent sentences, has he found ways to build and add and accumulate, no matter the activity or toy?

In between my mowing grass and my analysis of a preschooler, our struggle to subtract, and the potential rewards for doing so, have never been far from my thoughts. As an undergraduate, I majored in civil engineering, which, with its focus on creating buildings and bridges, is the professional version of Ezra playing with blocks. Ironically, as I learned the science and math of adding physical infrastructure, the sheer amount of content made me see the value of mental subtraction.

After college, I spent a few years building schools in New Jersey. I saw how removing floor tiles could create easier-to-clean, and therefore healthier, learning environments. I saw how simplified construction schedules let milestones stick out. I saw how reassigning a redundant project executive made our office run more smoothly. And I saw how rare these subtractions were.

My thinking about less really took off when I became a professor. Not since I mowed grass had I gotten paid to think freely. And now I had some new tools to help me usefully think about less.

Like most other professors, I happily devote my work life to creating and sharing (and sometimes subtracting) knowledge. Unlike most other professors, I am relatively untethered to a single discipline. My official title mentions engineering, architecture, and business, but many of my closest collaborators identify as behavioral scientists. My discipline-spanning role means I have more meetings to subtract from my calendar and more emails to filter from my in-box. But such inconveniences are a small price to pay for the one-of-a-kind network whose ideas and work you will meet in *Subtract*.

It is my academic trespassing that has honed my thinking about subtraction. Lawn-mowing musings deservedly confined to my own mind have been honed over the last decade into evidence that I feel excited—and obligated—to share with as many people as possible.

Before we examine that evidence, we need to know what we are looking for. Here is the conceptual distinction that advanced my thinking—a few thousand hours of me trying to get somewhere, all condensed into two paragraphs for you:

The breakthrough came when I figured out that what I am interested in is not simplicity, or elegance, or any other form of "less is more." Subtracting is an action. Less is an end state. Sometimes less results from subtraction; other times, less results from not doing anything. There is a world of difference between the two types of less, and it is only by subtraction that we can get to the much rarer and more rewarding type.

In other words, subtraction is the act of getting to less, but it is not the same as doing less. In fact, getting to less often means doing, or at least thinking, more. Removing a freeway is far more challenging than leaving it alone or than not building it in the first place. As my team would find in our studies, mental removal requires more effort too. So, subtractors need not be minimalists, laid-back, anti-technology, or possessed of any other philosophy that owes some of its popularity to its ease. In fact, when we mix up these other philosophies with subtraction, we don't see taking away as an option, and we discount the hard work needed to make it happen.

With my thinking clarified, my team embarked on tens of thousands of hours of research. We experimented, discussed, wrote, presented, and repeated. And we discovered that humans overlook subtraction. People don't think of the other kind of change, even when subtraction is obviously the better option.

The next question I had to ask was: Why? And then, how can we all get better at seeing subtraction?

4.

I'm not the first to notice the power of taking away. There is plenty of advice that works because it brings us to less: computer scientist Cal Newport preaches digital minimalism; chef Jamie Oliver distills recipes down to five ingredients; and the tidying savant Marie Kondo declutters homes. Each of these gurus guides us to specific ways we can subtract to improve. And their counterintuitive advice brings joy.

But why does this advice remain surprising? And why did I need to read three different books to fix the same basic problem in my computing, cooking, and cleaning? It's been five centuries since Da Vinci defined perfection as when there is nothing left to take away; seven centuries since William of Ockham noted that it is "in vain to do with more what can be done with less," and two and a half millennia since Lao Tzu advised: "To attain knowledge, add things every day. To attain wisdom, subtract things every day."

I have learned plenty from these new and old prophets of subtraction. But the main takeaway is that they are the exceptions proving the rule: their advice endures because we are still neglecting subtraction.

While the consequences of this neglect are most apparent in our surroundings, those visible effects stem from how we think. Ralph Waldo Emerson poetically links our thoughts and our physical world in his essay "Nature":

> Observe the ideas of the present day . . . see how timber, brick, lime, and stone have flown into convenient shape, obedient to the master idea reigning in the minds of many persons . . . It follows, of course, that the least enlargement of ideas would cause the most striking changes of external things.

William James, one of the founders of psychology, observed essentially the same thing in the other direction, describing in *The Principles of Psychology* how our homes and other material things become part of our personality.

That's why I trespass across isolated academic islands—from design to behavior, engineering to psychology, architecture to business and policy. To understand the relationship between thinking, the creations our thinking inspires, the thinking that results, the things that are created, and so on . . .

This symbiosis between our external and internal worlds is potent. It's why Sue Bierman didn't just remove a freeway, she helped San Franciscans rethink the relationship between vehicles, people, and their town. The convergence of things and ideas is why Leo Robinson didn't just spark divestment, he helped people in America see they had brothers and sisters in

South Africa. And by going from ideas to things, Elinor Ostrom didn't just remove a wrong idea about common resources, she transformed environments from the forests of Indonesia to the fisheries of Cape Cod.

If there was ever a time we needed all our Elinors, Leos, and Sues, it's now. The COVID-19 pandemic, at horrific cost, has given us a singular chance for change. It has forced us to rethink our daily schedules, our streets and cities, and our society. In our climate commons, the pandemic has so changed travel and consumption that carbon dioxide emissions are veering downward for the first time. What will we do with this blip in our selfish herding? Will we choose to lock in some of the pandemic-imposed less? At the very least, can we agree to permanently subtract the conferences and commuting we haven't missed? Then there is the latest awakening to the sinister endurance of systemic racism in the United States. What will we do with our reckoning? African Americans are three times as likely as whites to get COVID-19. Will we be content with a few more African American public health officials? Or will we use this chance for change to subtract structural racism—like the redlining legacy that confines African Americans to neighborhoods lacking options for healthy food and exercise?

Now as ever, there is no single approach, not to change our schedules or our minds, and not to improve our cities or our political systems. The project of this book, then, is to help us tap into the power of the other kind of change—by diagnosing (Part I) and then treating (Part II) our subtraction neglect.

PART I

Seeing More

1

Overlooking Less

Legos, the Lab, and Beyond

1.

An epiphany in my thinking about less came when Ezra and I were building a bridge out of Legos. Because the support towers were different heights, we couldn't span them, so I reached behind me to grab a block to add to the shorter tower. As I turned back toward the soon-to-be bridge, three-year-old Ezra was removing a block from the taller tower. My impulse had been to add to the short support, and in that moment, I realized it was wrong: taking away from the tall support was a faster and more efficient way to create a level bridge.

Since I had become a professor, I had been trying to convert my interest in less into something I could study instead of just ponder. From the start, I studied ways that buildings and cities might use less energy—and therefore produce fewer climate-changing emissions. I studied architecture and urban design, the people using it, and the people designing it. Over time, I had homed in on the designers, finding that, even when it leads to suboptimal things, designers use mental shortcuts: anchoring on irrelevant numbers, unthinkingly accepting default choices, and being swayed by examples. Still, I could never quite get from studying buildings and cities to studying less itself.

Ezra's encounter with Legos took my applied thinking about design to a more basic level. Here, right in my living room, was a relatively simple situation that could be changed by adding to it and by taking away from it. And when Ezra's choice caught me by surprise, it made me realize that, whereas less is an end state, subtracting is the act of getting there.

Not only did Ezra's bridge shift my focus from less to subtraction, it gave me a convincing way to share and test my epiphany. So I began carrying around a replica of Ezra's bridge. I tried it out on unsuspecting students who came to meet with me, checking whether they would subtract, like Ezra, or add, like I had. All the students added.

I also brought the Lego bridge to meetings with professors and one of them was Gabrielle Adams, who had appointments in public policy and psychology. Gabe and I were recruited to the University of Virginia at the same time. I had taken the job for the chance to work with scholars who studied human behavior in contexts outside design and Gabe fit the bill, having made contributions to workplace politics, ethical transgressions, apologies and forgiveness, and more. Impressed with her track record—and having commiserated over redundant new employee orientations and nocturnal babies—I had been pitching Gabe ways we might work together ever since.

I had tried linking my interest in less to Gabe's research: "removing the office jerk can improve workplace politics." I had tried linking less to Gabe's care for the environment: "designs with fewer materials can allow human progress without plundering natural resources." I had even tried linking less to popular trends: "Tim Ferriss's *4-Hour Workweek* says that we should have skipped those new employee orientations." Gabe always heard me out, but she didn't discern an idea that was a worthy use of our time. Taking on unclear ideas isn't what got Gabe named by *Poets&Quants* as a "best professor under 40"—when she was still in her twenties.

Thankfully, Ezra (who has proven just as unlikely to subtract as grown-ups) had given me a hands-on demonstration for my next meeting with Gabe. I pulled the Lego blocks out of my book bag, set them on Gabe's desk, and asked her to make the bridge.

Because of her intellect and because of our prior conversations about less, I suspected Gabe might see right through the bridge challenge. But she was like the others, and like me. She added a block to the shorter column to make the bridge.

Excitedly, I told Gabe how Ezra had removed a block—and that's when it clicked for her. Her response gave me the language to bring countless others up to speed, without them walking around in circles behind a lawn mower and then spending hours playing Legos with a toddler. She said, "Oh. So, you're wondering whether we neglect subtraction as a way to change things?"

That sounded right to me.

2.

Once Gabe figured out the question I was asking, she was on board, and she convinced Ben Converse, another psychology and public policy professor, to join us. Knowing we'd eventually need to study *why* people underuse subtraction, Gabe wanted Ben's expertise in the basic thought processes of human judgment and decision-making.

Gabe teaches graduate courses in experimental design. Ben lives it. He and his partner, who is also a psychology professor, met at a seminar on experimental design.

Many eager parents, and I suspect all married psychology professors, do the "marshmallow test" of delayed gratification on their preschool-aged kids. In the original version of the test, kids were given one marshmallow and told they could have another if they hadn't eaten the first one after a few minutes. Some kids waited patiently. Others gobbled up the first marshmallow and sacrificed the reward. Follow-up studies on the same kids showed that preschoolers who waited for a second marshmallow turned into teenagers who scored higher on the SATs. What causes this correlation is unclear, so don't stop parenting no matter what your preschooler does with the marshmallow. What you need to know is that Ben not only ran the marshmallow test on his child, he also had his partner blindly replicate it, to be sure. That's the kind of researcher you want on your team; just be prepared to do a lot of studies.

My first studies with Ben and Gabe used Legos. We had research assistants recruit passersby on campus. The assistants escorted willing participants to a workstation where a Lego structure sat on a small desk with more blocks scattered in a pile off to the side. Participants worked on structures of either eight or ten Lego blocks, each arranged on an eight-by-eight flat base.

Each participant changed the structure as they wished and then returned it to the research assistant, who counted the total number of blocks added, removed, and/or moved. Among the transformed structures, just 12 percent had fewer blocks than the original.

Now, maybe what we were observing was specific to Legos. We wanted to know whether this apparent underuse of subtraction extended to other situations. Where, if anywhere, did the behavior stop?

We asked people to change random "loops" of musical notes. They were about three times more likely to add notes than to take them away. It was roughly this same three-to-one ratio when we tasked people with improving a piece of writing. We asked participants to transform a five-ingredient soup. Two out of ninety participants subtracted ingredients.

Still, how could we be sure that we hadn't created situations that discouraged subtraction? In our writing study, for example, maybe we had provided initial writing samples that were missing key information, and therefore required adding. While I knew we had not intentionally stacked the deck against subtraction, hanging out with psychologists had me leery of my subconscious.

One way to be sure we were not inadvertently favoring addition was to outsource the creation of these situations. We tried the Lego scenario again, this time with initial structures that were created by a randomized process. Only one out of sixty participants subtracted. We had the initial Lego structures created independently by other participants. Only 5 percent took away to improve these independently produced structures. We removed ourselves from the writing scenario, too, by having another set of participants create the initial piece of writing. These participants summarized an article (about the discovery of King Richard's bones beneath a parking lot), and the participant-generated summary then became what the other set of participants were asked to improve. Only 14 percent took away to make a shorter summary.

People were overwhelmingly adding, regardless of whether the original situation was created by my team, by a random process, or by others.

We then created a situation we were sure would inspire subtraction, asking participants to improve an itinerary for a day spent in Washington, D.C. Over the course of fourteen consecutive hours, this itinerary had the participants visiting the

White House, Capitol Building, Washington National Cathedral, United States National Arboretum, the Old Post Office, and Ford's Theatre; with additional stops to pay their respects at the Lincoln, World War II, and Vietnam Veterans Memorials; and, rounding out the itinerary, a museum visit, shopping, and lunch at a five-star bistro. Travel time alone between all these stops would exceed two hours, and that is before considering D.C. traffic.

Participants saw this original itinerary in two sections: "Morning: 8:00 a.m.–3:00 p.m." and "Afternoon/Evening: 3:00 p.m.–10:00 p.m." Using a drag-and-drop interface, participants could change their itinerary by rearranging, adding, and subtracting activities. Only one in four participants removed activities from the packed original.

Building, writing, cooking, scheduling—our findings showed that adding more than subtracting is widespread. Our next question was: Could we generalize what we were observing? All else being equal, would people add more than they subtract?

To test this, we had to design a study that was context-free, meaning any behavior we observed could not be explained by something familiar about the study. We wanted to observe how people changed a situation with which they had no prior experience, and therefore no habits or inclinations to bring to the task. Such a design would go a long way toward showing that what we had observed in Legos, soup, and essays applied to everything else too.

When Gabe and I added Ben to our team, we also picked

up a postdoctoral student who was working with him at the time. Andy Hales is now a professor at the University of Mississippi, who, when he is not working on subtraction, studies ostracism and best practices for replicability. Andy is laid-back, except when it comes to devising, conducting, and picking apart his research. For that, he turns into a super-caffeinated version of Ben. More good fortune for me.

After helping with some of our early studies, Andy had become the driving force behind the quest for the context-free study. His iterating led to six different grid patterns like the one below. Pretend you are a participant and try this one.

Your task is to make the patterns on the left and right sides of the dark middle line match each other, so that if either one were lifted up and placed directly on top of the other, there would be a perfect fit. Your challenge is to do so by making the fewest possible changes.

There are two best responses. One is to add four shaded blocks to the left side. The other equally simple and correct response is to remove four shaded blocks from the right side.

Even in this decontextualized case, participants were more likely to add than subtract squares. And again, it wasn't even close. Just 20 percent of participants were more likely to subtract to transform the grids.

The evidence from Andy's grids was clear: it wasn't just habit that explained all this adding. Nor were people adding more just because they like the things being added or subtracted. If people see a single Lego block in a larger structure or tomatoes

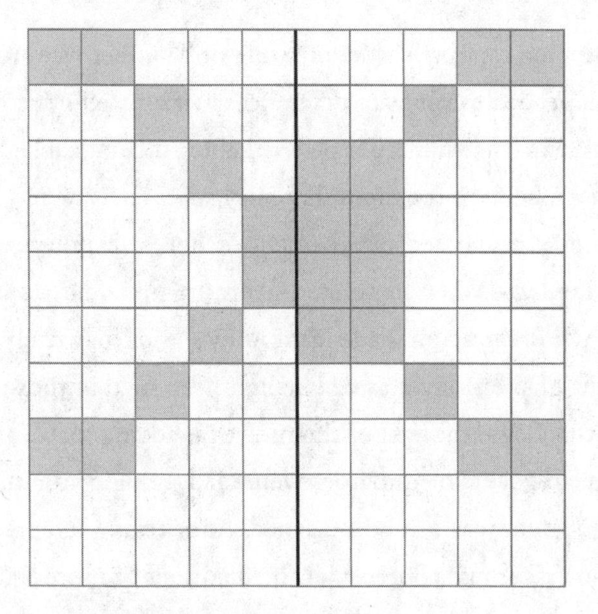

Figure 1: One of Andy's grid patterns

in a soup as inherently valuable, they may resist taking them away, even if doing so would improve the structure or soup. But there is no inherent value to squares on a computer screen. The adding on Andy's grids could not be explained by love of digital squares.

Also thanks to Andy's grids, we learned that all of the adding was not due to varying effort required to add and subtract. It can be kind of hard to pull apart Legos. And, while the soup recipes were just a list on a screen, perhaps participants were imagining how much of a hassle it would be to physically remove tomatoes from the rest of the mixture. On the grids, however, it was the same click of a mouse or touch of a screen whether turning a square from shaded to white or the other way around. We still needed to consider differences in *mental* effort required to add and to take away, but physical effort did not explain why people added far more often than they subtracted.

When I presented our work to other researchers, right about here was when someone usually pointed out that participants who subtracted shade from Andy's grids may really have thought of themselves as *adding* lightness, or that those who took away Legos thought of themselves as adding space, or that those who got rid of tomatoes were really adding yumminess. This question was a valid nuisance. After trying to appeal to common sense and consensually held notions of more and less, we began gathering evidence.

One way to know what people are thinking is to ask. After completing the grid tasks, we asked participants to indicate their approach to transforming the grids: either "I added squares until they were symmetrical," or, "I removed squares until they were symmetrical." These self-reported responses confirmed that people were thinking about adding as adding and subtracting as subtracting.

The evidence was, well, adding up.

The beauty of well-designed studies—especially when there are lots of them getting at the same thing—is that their lessons can extend well beyond the studies themselves. I couldn't help but jump to the implications of what we were seeing.

One valid interpretation of our results, I reasoned to Ben, was that if subtracting is as useful as addition yet is used less often, then there is untapped potential: people are consistently neglecting a basic way to make change. Such neglect would have something to say about everything from the difficulty San Franciscans had removing their freeway, to our tendencies to clutter our homes, schedules, and minds.

But we weren't yet ready for implications. As Ben reminded me, "We need to go from *can* we believe that people neglect subtraction to *must* we believe that is the case."

Research requires a unique blend of doubt—to avoid jumping prematurely to "we must believe"—and confidence, to think that you can actually get there. Ben is a reliable source of doubt. For the confidence, we were getting external validation, albeit mixed with prodding. Ben brought some of our initial findings to a judgment and decision-making conference in Seattle, and whatever transpired in his conversations there, he returned more assured that we at least had an "interesting phenomenon" on our hands. Around the same time, I presented at Princeton University and got a chance to ask one of my role models for her thoughts. She started with, "It's really neat," and then, as Ben had conveyed in his "interesting phenomenon" remark, she continued with a caveat that made clear we had more to do: "It's a nice philosophical question."

For me, there's nothing better than a nice philosophical question. My favorite consequence of discovering our interesting phenomenon had been discussions with all sorts of people about how it might apply in the things they cared about changing. At the same time, our work hadn't yet answered a major question. Fine, we don't subtract as much as we add. That knowledge is an interesting phenomenon. But *why*? Knowing why would tell us whether all of our adding was causing us to miss out. And, if so, knowing why would surely reveal ways to stop missing out, by finding less more often. Now that would be useful.

One explanation for what we had seen in our studies nagged at me: perhaps subtraction wasn't as good as addition, subjectively. Maybe people just like Legos with more blocks, essays with more words, and grids with more shaded areas. If the person who added ingredients preferred their soup to have a more complex taste, or if the person who added another museum in Washington, D.C., preferred packed itineraries, then the adder made the right choice. The adder may not necessarily even prefer the results, they might simply like the act. Maybe we choose adding because we like things better that we have built ourselves—the IKEA effect. Perhaps we choose adding because to take away is to admit that previous additions are sunk costs. Or maybe we choose not to subtract, because we assume that, if something exists, there's a good reason for it. Or because losses loom larger than gains. Sure, getting rid of a wrong theory, an eyesore freeway, or apartheid is not a loss. But, as we'll see in chapter 5, it's easy to mistake less for loss.

If we choose to add, for whatever reason, then our "interesting phenomenon" isn't necessarily a problem. But what if we are not even considering subtraction? In that case, if we aren't seeing the possibility at all, then we are definitely missing out.

3.

My team needed to go from "can we believe" to "must we believe." Did people willingly choose hectic travel itineraries, or did they fail to even imagine free time as an option?

In Ben's judgment and decision-making circles, this was a question of mental "accessibility." I think of mental accessibility as similar to the physical accessibility in Ezra's toy closet, where books and art supplies occupy the shelves at his eye-level, while boxing gloves and slingshots occupy higher shelves. In Ezra's closet, accessibility affects how often he uses different toys. In our brains, accessibility affects how often we bring stored ideas to bear on our surroundings.

Accessibility promotes mental efficiency. An idea we used yesterday is more apt to be useful today than one we last used twenty years ago. Our brain stores both ideas, but yesterday's is easier to reach. Accessibility can also lead us astray, though; we underestimate the danger of car travel and overestimate the risk of air travel, because plane crashes are more memorable and therefore more accessible. In this case, the idea that's easy to reach can cause us to choose the less safe way to travel.

Whether traveling to see the grandparents or transforming Andy's grids, our journeys don't begin with a blank mental slate. Sure, how we change a situation depends on our conscious choices, but those choices depend on what comes to mind quickly and easily; just as Ezra is more likely to use the books at eye-level, we are more likely to use easy-to-find ideas. The conjecture in Ben's professional circles, and at his dinner table, was that subtraction neglect was a symptom of adding being the more accessible kind of change.

We could test that.

If adding was more accessible than subtracting, we hypothesized that three approaches would reduce this difference:

- searching our minds more deeply for ways to change the situation;
- bringing subtraction specifically to mind; and
- devoting more mental bandwidth to the change effort.

One way to induce deeper mental searches is through repetition, which forces us to think beyond the first idea that comes to mind. We returned to Andy's grid patterns to test this. What if we had participants come up with multiple ways to solve the grids? Would they eventually think of removing shaded areas—and then, might they choose that change?

In this experiment, we required some participants to do three "practice" runs of the grid task before the "official" fourth test. Sure enough, participants were more likely to subtract on the official test than they were on the practice runs. And they were more likely to subtract on the official test than those who hadn't done any practice runs. Once people thought of the option that removes grid blocks, they were more likely to choose it as their favorite. Deeper searches gave people more opportunity to find the subtractive solution—and to realize that they liked it.

Evidence from this and similar studies brought me back to one of the earliest talks I had given about our research, back when all I could say was that people don't like dismantling Legos. In the post-talk discussion, an architecture professor had offered that one of the ways he tries to help his students overcome blind spots, whether related to subtraction or not, is by requiring them to come up with five, ten, or even fifty conceptual changes to the problem at hand. Only then are students allowed to choose their preferred change to develop more fully. Our newest evidence matched his experience.

Asking people to repeat the problem could help them solve it with subtraction. That evidence hints at some of the subtracting skills we will focus on in the latter half of this book. It also got our team closer to "must we believe" evidence that neglecting subtraction is harmful. It's not just that we accept objectively worse results because we subjectively like adding, it's that we fail to even consider subtraction in the first place.

Our second test of accessibility was to bring subtraction specifically to mind, overriding how our brains have stored the two kinds of change. We designed the tasks for these experiments so that taking away was, in retrospect, the clear best choice; if subtraction were accessible, participants should choose it.

By this time, Ben's son was at the tail end of his superhero phase. His idle action figures and Ezra's extra Legos led to our storm trooper sandwich study. In it, we challenged participants to modify a sandwich-like structure made from Legos so that it was strong enough and high enough to hold a masonry brick above the head of a storm trooper figurine.

Each participant received the structure shown below, consisting of parallel horizontal Lego panels connected by a vertical column that narrowed to only one block wide where it connected to the top panel.

Figure 2: Storm trooper in a Lego sandwich showed that people overlook subtraction, in one specific context. (Photo by Elliott Prpich)

We asked participants to:

"Improve this project so that it can hold a brick above the storm trooper's head without collapsing."

And we offered an incentive:

"You will earn one dollar if you successfully complete this task. Each piece you add costs ten cents."

Try it.

The best solution is to remove the single block forming the thin part of the column. The top panel can then be attached to the larger section of the column, which stabilizes the structure and still leaves enough clearance to avoid the storm trooper getting squashed by the masonry brick.

Subtracting one block was the fastest way to solve the problem. Plus, only subtracting allowed participants to earn the full dollar.

And yet participants were still more likely to add than subtract. This was evidence that people add to their detriment—at least when trying to modify a Lego structure so that it can hold a brick safely above the head of a storm trooper.

But the specifics of the Lego structure did not matter because here we were not interested in exactly how many people thought of subtracting. We wanted to know whether the rates of subtracting would be higher among those for whom subtracting was brought to mind than among those for whom it was not.

To try to override the greater accessibility of adding, we would give some participants subtle reminders, or cues, that subtraction was an option. If those who received the cue subtracted more often, then that would indicate that those who didn't receive the cue were overlooking subtraction.

The experimenter said to all participants, "You will earn one dollar if you successfully complete this task. Each piece that you add costs ten cents." Participants randomly assigned to the cue condition heard one more instruction from the experimenter: "but removing pieces is free and costs nothing."

Those eight words in the instructions were the cue: the only difference between the groups.

In the no-cue group, 41 percent subtracted a block. In the cue group, 61 percent subtracted. Those who were cued took home an average of eighty-eight cents, 10 percent more than those who didn't get the cue. The simple and subtle eight-word cue showed people a profitable solution that they had otherwise been missing. It sure seemed like people who didn't receive the cue were missing the subtractive option not by choice but because they couldn't see it.

To be certain, we piled on more evidence with an experiment designed by Gabe, who was tiring of Ben's and my affection for Legos. In this experiment, all participants were asked to imagine themselves as the assistant manager of a miniature golf course. Participants saw an image of a hole from their course and were asked to "make a list of all the different ways that you might be able to improve the hole without spending a ton of money." Advising against "spending a ton of money" was one way we encouraged all participants to consider subtractive changes. We also designed the miniature golf hole to include opportunities for subtraction. Participants who wanted to make the hole more challenging could remove the corner bumper that golfers might use to carom around the corner, while participants who wanted to make the hole easier could remove the sand trap.

As with the storm trooper sandwich, participants were randomly assigned to receive a written cue, in this case: "Keep in mind that you could potentially add things to the hole as well as take them away."

This cue made both adding and subtracting accessible. Therefore, when the cue increased rates of subtraction (from 21 percent to 48 percent) but not addition, it told us that the additive part of the cue was redundant with ideas people were already considering, whereas the subtractive part of the cue brought new ideas to mind.

There was one more important wrinkle in the miniature golf experiment. There can be many goals for change, most often trying to aid the existing situation. Sue Bierman wanted to improve San Francisco's waterfront. Elinor Ostrom wanted to advance knowledge about how people manage common resources. Sometimes, though, the goals are to interfere with or even sabotage the existing situation, as when Leo Robinson worked to dismantle apartheid. To study subtraction neglect when the goal is sabotage, we ran a version of the miniature golf experiment in which participants were asked to play a friendly prank on their rival's course, transforming a hole to "make it worse." Whether the goal was to "make it better" or to "make it worse," the cue that participants could add or subtract increased the percentage who subtracted (but not added). These results imply that the same oversight that keeps us from better cities also holds when we try to disrupt racism.

At this point, I was as convinced as I needed to be that we add more than we subtract, that doing so causes us to miss out on good options, sometimes because we don't even think of taking away. Hopefully, you're convinced too. But, as I did, let's take it to an even higher "must we believe" threshold.

After miniature golf, my team still had the bandwidth theory of accessibility to work with. Our mental processing power faces competing demands; this is why we shouldn't drive distracted, why we can't engage in two conversations at once, and why being poor and forced to worry about money leaves less brainpower to devote to other areas of life. In the next chapter, we'll see how our brains subconsciously subtract to protect our precious mental bandwidth, and in chapter 8, we'll learn how we can consciously do the same. But for now, to confirm accessibility, we just wanted to change bandwidth and observe what happened. In theory, more bandwidth should have made subtraction more accessible, and vice versa.

Predictably, despite being thousands of hours into research on neglecting subtraction, our first instinct was to add to our participants' bandwidth. We thought about giving people more time to think. Could we randomly assign some participants to a modified version of our instructions, asking them to take five minutes to ponder their options before settling on their favorite?

If we tried that, we couldn't be sure they would really use the time to think about our study and not something else.

What if we took away bandwidth? Would that make subtracting even less accessible than it normally is? There were a couple of interventions known to induce cognitive load and therefore reduce bandwidth. And we now had a series of Andy's grids for which subtracting was not just an equally good change, it was obviously better, if people thought of it. For these grids, the participants' task was to make each grid symmetrical from left to right *and* from top to bottom, using the fewest clicks possible. Symmetry could be achieved with additions to the three empty quadrants, but fewest-clicks symmetry could only be achieved with subtractions from the one populated quadrant. Participants could make the correct change only if they thought to take away.

In an in-person version of this experiment, we instructed participants in the control group to complete the grid tasks while sitting naturally. To subtract bandwidth from the other group of participants, we instructed them to do the following:

"Please move your head around in a circle. (Focus on moving your chin in small circles, in whichever direction is easiest for you.) You should move in a controlled, continuous fashion (nothing wild). Please continue to execute the movement during the entire set of patterns, starting now."

Go ahead and try it if you like. Just remember to stop before reading on.

In online versions of the experiment, we sent a stream of

numbers across the participants' displays as they worked on the grids. Participants in the control condition were told to ignore the scrolling numbers, whereas participants in the reduced bandwidth condition were instructed to press the F key on the keyboard every time the number 5 came across the screen.

Across all these studies, with multiple grid patterns and more than 1,500 participants, less bandwidth meant less subtracting. Along with cues and deeper searches, the bandwidth affect gave us a trifecta of evidence that the first reason we neglect subtraction is because we don't even think of it.

Like my meeting with Gabe about Ezra's bridge, our meeting to discuss the bandwidth results has stuck with me. It had taken thousands of hours of collective conjecture and discussion, reading and analysis. At the time, there were (ironically) more than 250 subfolders and 1,700 files in our shared "subtract to improve" folder. Andy finally admitted that he now "must believe" in accessibility. Ben agreed, smiling so broadly that we could see white teeth through his winter beard. And Gabe, conferencing in from a trip to California, reminded us of the even bigger picture: we also had to believe that accessible adding deprives us all, because when we do finally see subtraction, we like what we see.

4.

The advantage of studying subtraction in Legos, miniature golf holes, and Andy's grids is that those experimental contexts allowed us to tightly control the conditions. We could remove variability and therefore rule out other explanations for subtraction neglect. We could make it so that the only difference between two groups solving grid patterns on their computer screens was that one group was asked to also report the scrolling numbers while doing so. When subtraction was less likely for that group, we could be pretty sure it was thanks to the reduced bandwidth.

Real-world change is messier than our purposefully narrow studies. Some San Franciscans probably found it hard to imagine (literally and figuratively) getting rid of the Embarcadero Freeway. But lack of accessibility, not even thinking of subtraction as an option, is not a full explanation.

There are surely other psychological forces, and cultural, and economic ones too, that explain why subtracting is neglected—and why it took decades and an earthquake for San Francisco's eyesore to be removed. My team's findings therefore gave rise to all sorts of next-level questions. To what extent are we conditioned to add through Lego-building childhoods and home-adding adulthoods? Are genetics a factor? Do material wealth and scarcity make subtraction more or less likely?

As we're about to see, our subtraction neglect has deep and tangled roots in our nature and nurture. These roots span scientific disciplines and professional norms, from our uncivilized history to our modern affinity for growth. The more we know about all of these forces, the better we can become at finding the delight of less.

2

The Biology of More

Our Adding Instincts

1.

Bowerbirds spend lots of time and energy building elaborate, seemingly useless nests. Depending on the species, the male might place sticks around a sapling to create a round hut or assemble walls of vertical sticks into a shotgun shack. Like Realtors staging a home, they decorate these bowers using everything from shells, flowers, and berries to coins, nails, and rifle shells. When the building and staging is finished, female bowerbirds tour the available bowers, even returning to their favorites for another round of inspection. Evidently, each female bowerbird just needs to decide where she will settle down.

Once the female bowerbird chooses her favorite bower, however, and mates with the builder, she goes off to build another nest. This nest is where the female shelters and raises the next generation of bowerbirds. The bowers are never used as nests in the traditional sense. The whole purpose of these avian palaces, and all the work that goes into them, is to provide visible proof that the builder has good genes.

Remember those records of my subtraction research, those 250 subfolders and 1,700 files I told you about in the last chapter? Sure, a few of these folders and files are necessary for record-keeping. If someone wishes to replicate our studies, we want to be able to share every last detail about what we did. And maybe we'll be able to copy and paste from some of the old files. But we probably don't need to preserve drafts one through eighteen of our first paper for posterity (I wish I were exaggerating). Perhaps you are more judicious in your electronic filing, but surely I'm not the only one who has kept an unnecessary section in a report, or displayed a book I never intend to read.

Could my excess archiving be something innate, as with the bower building? An evolutionary advantage gone awry? Signaling fitness is, after all, one of the basic biological reasons for behavior. Bowers signal fitness; a male bowerbird that is good at assembling and staging stick structures might also be in good general health. But what about me adding yet another electronic folder?

In 1959, Harvard University psychologist Robert W. White took a step toward connecting file folders with evolution. In a paper that has been cited more than ten thousand times, White described our "intrinsic need to deal with our environments"—not just for survival but to avoid feeling helpless. White defined his key idea with one word, *competence,* meaning how well we feel we are dealing with our world. In 1977, the Stanford University psychologist Albert Bandura extended White's idea, concluding that one way we meet our intrinsic need to feel competent is by successful completion of tasks. Our biological need to deal with our world is also why it feels good to check items off of a list (and to complete yet another draft of a paper).

Why would our intrinsic need to feel competent work against subtracting? After all, Ezra learns about his world when he takes away Legos just as well as when he adds them. It's true; we can *develop* competence just as well by subtracting. The problem is that it can be harder to *show* competence by subtracting.

When we transform things from how they were to how we want them to be, we need proof—to show mates, competitors, and ourselves. Adding a freeway or file folder shows the world what we did. But just as the Embarcadero Freeway has disappeared from San Francisco's waterfront, there is no proof of the few shared file folders I managed to prune. No matter how beneficial an act of subtraction is, it's not likely to leave as much evidence of what we've done.

In the last chapter, we learned the first way humans miss out on less: because it does not even come to mind. To decide that less is more, we need to see it as an option in the first place, and often we don't. But oversight isn't the only way we neglect subtraction. As we're about to see, our biology (this chapter), our cultures (chapter 3), and our economies (chapter 4) not only contribute to this oversight, they also lead us to reject perfectly good subtractions even when we do think of them.

Biological, cultural, and economic forces overlap, of course. Adding the freeway showed competence, which stimulated automobile culture in San Francisco, which changed the economic costs and benefits for future development in the city. All these overlapping forces create an unsolvable "this causes that" puzzle. But the variety is good for us. By understanding our adding from many angles, we gain versatile skills for finding less.

So, let's first examine adding as biologists. Evolutionary forces that help explain our skewed approach to change are deeply ingrained. They have allowed nest-adding bowerbirds, freeway-adding humans, and all other living things to stay alive and pass along their genes. My research team may have created those 250 subfolders, but evolution created my research team, and evolution created my research team with a desire to show competence.

2.

The ancestry company was delicate with the news: I was "less than 4 percent" genetically Neanderthal, which at least sounded better than "more than 3 percent."

I was an early adopter of direct-to-consumer genetic ancestry testing, one of the first people to spit into a plastic test tube, mark my name on it, and send it off in the preaddressed and postage-paid envelope. I now receive a steady stream of updates on my heritage from the company that connects the genetic dots. Most of the time, the updates are because new saliva senders turn out to be fourth or fifth cousins of mine.

One memorable update, though, brought the news that I am part Neanderthal. Along with my "less than 4 percent" share, the update emphasized that nearly everyone lacking recent roots in sub-Saharan Africa is part Neanderthal, and that the mating between humans and Neanderthals is thought to have occurred a long time back, about forty thousand years ago, and across an ocean, in what is now Europe and Asia.

It turns out that forty thousand years ago was a crucial time in human history, especially for our interest in adding and subtracting. It was the beginning of the transition to "behavioral modernity," as our ancestors developed new abilities to think about things that were not physically present. For the first time, humans could imagine the future implications of less and more.

Before behavioral modernity, subtracting was not a conscious choice, and neither was adding. Since humans could not even imagine a new situation, any changes could only have been due to instincts.

To examine how biological forces might favor adding, then, we need to start here.

At the dawn of behavioral modernity, survival meant constantly finding food. This was such an all-consuming task that it defined humanity, literally. Humans forty thousand years ago were all hunter-gatherers, and it had always been that way. For the previous one hundred thousand years or so, my ancestors had hunted and gathered their way from eastern Africa across the Red Sea into the southeastern part of the Arabian Peninsula and then into the Middle East and Central Asia, where finding food meant digging for wild eggplant and hunting mammoths with stone weapons.

For many of us, life no longer revolves around a quest for food, and health can even depend on eating less. But we still enjoy eating. Across our long history of chasing calories, eating helped us survive and pass on our genes. Much as I struggle to override it, this evolutionary instinct is what I blame whenever I polish off an entire bag of chips even though I am already full.

But what about when I store the free but too-small hat that came with Ezra's new skateboard? An uncertain food supply characterized most of human evolution, but not an uncertain hat supply. Is evolution to blame when I add Facebook friends who I have never met? Surely evolution cannot explain adding turned deadly? Can it?

On March 21, 1947, Homer Collyer's body was discovered in his four-story brownstone in Harlem, New York. Police found him in a tattered bathrobe, doubled over at the waist, with his head of long and matted gray hair resting on his knees.

Homer's housemate, caretaker, and younger brother, Langley, was nowhere to be found. One rumor, that Langley Collyer had been seen boarding a bus bound for Atlantic City, sparked a manhunt along the New Jersey shore. Police eventually searched nine different states looking for Langley, with no success.

The Collyer brothers had grown up as well-to-do sons of an opera singer and a Manhattan gynecologist. Both had earned degrees from Columbia University. The brothers lived relatively normal lives until 1932, when Homer had a stroke that left him blind. Langley quit his job to care for Homer.

According to Langley, he fed and bathed his brother, and even read Homer classic literature and played him piano sonatas. Langley hoped that he could cure Homer's blindness through rest and a strict diet, which included one hundred oranges' worth of vitamin C every week. But Homer never regained his sight, and he eventually became paralyzed from inflamed joints and muscles, which could have been treated if he had been taken to a doctor.

When police found Homer's body, they also found decades worth of adding. There were newspapers and magazines, which Langley was saving, so that Homer could catch up on the news when the oranges took effect. The brothers, who were childless, had stockpiled baby carriages and small chairs. They had acquired the top of a horse-drawn carriage, the chassis of a Model T, and fourteen pianos, all of which they stored in the brownstone. There was so much stuff that it piled to the ceiling.

Through all the stuff, Langley built a three-dimensional maze of tunnels. Like hamsters, he and Homer lived in nest-like openings in the results of their adding.

In the weeks after Homer's body was discovered, as the

hunt for Langley continued, gawkers came from near and far to watch the procession of things being removed from the brothers' brownstone. Finally, more than two weeks after they had found Homer, police found Langley. His decomposing body was ten feet away from where they found Homer.

From there, it became clear what had happened. Langley had died *before* Homer, from asphyxiation. Police reasoned that Langley must have been crawling through one of his tunnels when it collapsed and he was crushed. With no one to take care of him, Homer sat in his wheelchair and starved to death.

While neither of the Collyer brothers had children, it turns out that some of their adding is in all our genes.

A professor at the University of Michigan, Stephanie Preston knows more than anyone about what she calls "acquisitiveness," or how and why we get and keep things. To measure this, Preston created an object acquisition task that, like Andy's grids for my team, gives Preston a computer-based experimental approach that can be carefully controlled.

In the first part of Preston's task, participants see more than one hundred different objects, in random order, and one at a time. As each object appears on their screen, participants are asked whether they would like to acquire it virtually (the participants know they won't actually get the objects). All the objects are free, and participants can acquire as many, or as few, as they want.

The objects vary in their usefulness. There are bananas, coffee mugs, extension cords, and other items that many people would take for free. Others are less useful: empty two-liter bottles, used sticky notes, and, perhaps in a nod to the Collyer brothers, outdated newspapers.

Once they have made a choice about each object, participants are shown the full collection of everything they have added. A participant who has acquired seventy items, for example, is shown all seventy items together on the screen.

They are then encouraged to subtract objects.

First, participants are told that they may discard items from their collection, if they wish.

Then they are challenged to whittle down their collection so that it can fit into a shopping cart on the computer screen.

Finally, participants are asked to make the collection smaller still, so that it fits into one (virtual) paper grocery bag.

The goal is clear: everything needs to fit into one grocery bag. Participants even get real-time feedback, displayed on the computer screen, on whether they have subtracted enough stuff so that the rest fits. And yet, lots of participants fail to get down to a single bag. Many never make it past the shopping cart. People do not subtract their useless—not to mention imaginary—objects and, as a result, fail to accomplish the task.

Adding too much and then not subtracting enough may seem silly in experiments, but this same behavior turns sad when it is ruining real lives. Just as stress is linked to overeating, Preston has found that stress correlates with adding objects. In extreme cases, neglecting subtraction in the object decision task can be a sign of devastating anxiety and depression. In other words, those of us who choose to keep the free extension cord and coffee mug are exhibiting a milder version of the acquisitiveness that led to the demise of the Collyer brothers.

Like my team's earliest studies, Stephanie Preston's object decision task reveals a skewed approach to adding and subtracting. Some of Preston's other research suggests there may be biological roots to this behavior.

In this research, Preston's participants were kangaroo rats, which were useful subjects because they store food in the wild. Preston found that when the kangaroo rats had their piles of food stolen, they would again stockpile. This behavior seemed like no big deal when I first read about it. I get groceries when our stock is running low; kangaroo rats were just doing the same thing.

To appreciate why the rats' stockpiling is so significant, I had to remind myself that kangaroo rats are like humans before behavioral modernity. Rats don't think abstractly. Their adding cannot possibly be a result of thinking about how to change the current situation (no food stockpile) to a better one (food stockpile). Stockpiling must be an automatic response to environmental cues, an instinct that has helped the rats survive and pass down their genes.

This same stockpiling has been observed in other mammals and even in birds. When similar behavior is found across animals that evolved from a common ancestor hundreds of millions of years ago, as is the case with birds and mammals, that behavior is probably a biological instinct, one that extends through all of human history and beyond.

Our instinct to acquire food may also extend to our adding of other things. By having participants acquire things while hooked up to machines that show brain activity, neuroscientists have confirmed that food acquisition as well as other types of acquisition activates the same reward system in the brain: the mesolimbocortical pathway. This pathway runs from the outer layer of our brains, the cerebral cortex, which aligns our thoughts and actions with goals; into our midbrain structures that house emotional life; and deep into our ventricle tegmental area, the origin of dopamine pathways.

Because it connects these thinking and feeling parts of our brains, the mesolimbocortical pathway makes it pleasurable to eat. This same reward pathway can also be stimulated by drugs like cocaine and by website designs that keep us clicking and scrolling as we add Facebook friends, battle Twitter trolls, or buy books. For hoarders, even used sticky notes can provide a hit.

Even simple behaviors require coordination between many areas of the brain. That said, finding the role of a specific reward system does confirm just how deep-rooted some of our adding might be. And because our acquiring behavior maps to a key motivation system in our brains, it just might inhibit us from pursuing alternatives—like subtracting. This adding reward system, long helpful for food, is tough to turn off, no matter how unlikely I am to ever wear the free skateboarding hat.

Our instinct to acquire is, by its very nature, skewed toward more. But as behaviorally modern humans, we can resist an urge to stockpile bananas the day before we head out of town for two weeks. We can give away the must-have panini press that has become, in retrospect, no more useful than a stack of old newspapers for a blind Homer Collyer. The super-acquisitive even have treatment options, like cognitive behavioral therapy, that can help offset genetic predispositions. The more we learn about our adding instincts, the better we get at overriding them.

3.

Before he appreciated vegetables, we asked Ezra to eat three spoonfuls of them before having his dessert. One evening, when he was four, Ezra "forgot" about his three scoops of peas until he had already eaten his chocolate cake. Because no one likes to finish dinner on a vegetable, Ezra offered, "Tomorrow, I'll have five scoops."

Monica thought this was another negotiation ploy—Ezra trying to eliminate a scoop of peas from his two-day intake. She wasn't impressed with me when I told her that, based on some convincing research about our quantitative instincts, our son may have made an honest, and even predictable, error.

For going on four decades, Harvard professor Elizabeth Spelke has been studying the minds of infants and young children. Studying kids helps Spelke isolate the role of genetics, as opposed to education, in our learning and development.

In the research most relevant to Ezra's peas, and to our subtraction neglect, Spelke's participants were five- and six-year-olds selected because they knew numbers but hadn't yet learned to add or subtract. Spelke wanted to see whether children—and by extension the rest of us—possess an instinct for less and more.

The researchers in Spelke's lab tested this by giving each child participant three quantities as a word problem. Something like:

Rose has twenty-one candies.
She gets thirty more.
Ada has thirty-four candies.

Then, the researchers would ask the children, "Who has more candies?"

If you answered "Rose," nice job. You got it right. So, it turned out, did the kids who didn't know how to do arithmetic. They just had a feeling.

One might reasonably ask whether this spoke to a loophole in the study: maybe the kids did know arithmetic but didn't show their skills on the researchers' screening tests.

Spelke's study design ruled out explanations like this. Researchers varied the quantities in the word problems, such that some of the children received word problems for which the final candy difference was so small that intuition alone could not provide the answer. Whereas a child using math could get these close ones correct, a child relying on approximation would get fewer right answers the closer together the two choices were.

Questions like the following, for example, should be hard for those relying on intuition:

Rose has fourteen candies.
She gets nineteen more.
Ada has thirty-four candies.
Who has more?

To answer "Ada," one needs to know arithmetic.

For these close cases, children did no better at answering who had more candies than random chance would predict. The closer together the two final quantities were, the less likely the children were to correctly choose who ended up with the most candy.

The children were using intuition, not arithmetic. Just like Ezra when he offered to take five scoops of peas.

Studies like these from Spelke's lab were some of the first pieces of what is now convincing evidence that, even before we learn math, we perceive quantity. As with touch, sight, and smell, we have a sense for less and more.

In his book *The Number Sense,* Stanislas Dehaene weaves together the research findings, including his own, showing the pervasiveness of this innate sense. Kids who haven't learned math can do tasks that require intuition of relative quantity. So can adults in isolated Amazonian tribes, even though they have had no prior exposure to arithmetic. Mice, it turns out, sense less and more.

How might humans and other animals sense quantity without knowing mathematics—or even language, for that matter? Dehaene draws an analogy to the simplest version of a car odometer that is "just a cog wheel that advances by one notch for each additional mile." Given that such a simple device can record accumulated quantity, it's reasonable that living things might also have a mechanism that can do this, irrespective of high intelligence.

There are more than analogies to support this theory. Brain-imaging research shows that specific brain networks activate when we approximate quantity. These networks are closely linked to the networks known to help us sense space and time.

There is even a subject, "the Approximate Man," who had brain damage affecting his arithmetic networks, but not the ones he needed to sense quantity. When asked to add two and two, Approximate Man would give answers ranging from three to five. But, as Dehaene put it, "he never offers a result as absurd as nine." Approximate Man's math networks no longer worked, but, like Ezra, he could still intuit less and more.

Behaviorally modern humans combined their new capability for abstract thinking with this instinct for quantity. Both are needed to engrave notches in baboon bones and to paint dots on cave walls. From those earliest forms of counting, it was on to the numbers and math that allow us to separate concepts with close meanings, like thirty-three versus thirty-four candies. In this way, math enhanced our hardwired odometers.

When I first encountered the odometer analogy, it seemed unfair to subtraction. In my mind, at least, it implied that we only need to be able to sense more. Odometers only go up. Was this a flaw in the analogy, or does our instinct for quantity—like an odometer—actually favor more over less? My own research suggests that it's a bit of both.

To see why, and to learn what we can do about it, we need to go deeper into our instincts. When we don't have specific numbers to rely on, we judge the difference between large quantities as less than the difference between small ones. This is what we do with other senses too, like sound and taste. The first shake of salt on your vegetables changes the taste more than the tenth. Raising the headphone volume from one to two bars sounds like more change than raising it from eight to nine. The change we sense depends on the initial state.

An instinct for relative, as opposed to absolute, change would have been a helpful evolutionary behavior. For our roaming and hungry ancestors, the difference between eight and nine mammoths was far less important than the difference between one and two. A group of nine mammoths presented basically the same threat to survival and opportunity for some protein as a group of eight. Alternatively, trying to take down one mammoth with a handheld rock was a lot easier if its friend wasn't there.

What matters is not the absolute amount added, which is

the same in both comparisons (one mammoth). More useful is the relative amount added: 13 percent more mammoths when moving from eight to nine, versus 100 percent more mammoths when moving from one to two. When we rely on our instincts, the change between one and two is bigger than the change between eight and nine. The larger the quantity, the smaller the difference we perceive for each new unit of change.

Knowing how this sense for relative quantity works, we can ask the all-important questions: Does it favor more? Does it disadvantage less?

Let's return to the kids and candies. Children are less precise when using this instinct to estimate the result of eighty minus thirty than they are when using it to estimate twenty plus thirty, even though the result, in both cases, is fifty. If children arrive at fifty candies by *adding* thirty candies to their stash of twenty, their precision is based on their sense of thirty and twenty. If, on the other hand, children arrive at fifty candies by *subtracting* thirty candies from their stash of eighty, their precision is based on their sense of thirty and eighty.

The children sense thirty the same in both cases. But their sense of eighty will be less precise than their sense of twenty.

When we rely on approximation, our precision depends not on the final quantity but on the amount we used to get to the final quantity. This means that, for equal outcomes, addition should be more precise than subtraction. Maybe this is why "start small" is such well-received advice for making change. It's harder to imagine starting big.

I was struck by what our sixth sense might tell us about subtraction neglect. Was it harder to imagine what a city changed by a freeway removal would look like, or a racist system changed by divestment? Maybe, I thought, being worse at intuiting less is one reason we find it hard to think of subtracting as a possibility. Like our instinct to acquire, our instinct for quantity certainly doesn't draw us to less.

Three and a half centuries before Stanislas Dehaene, another French scientist, Blaise Pascal, was making the contributions that now have his name gracing a unit of pressure, a programming language, and a probability theory popular with actuaries and gamblers. But even the brilliant Pascal overlooked one type of less. In his aptly titled book, *Pensées* (*Thoughts*), he wrote, dismayed: "I know people who cannot understand that when you subtract four from zero what is left is zero."

While this statement seems odd to us, it was in line with the thinking at the time—a consensus with origins in the fact that one could not hunt fewer than zero mammoths or make fewer than zero counting notches on the baboon fibula. By the time of Pascal, so many brilliant minds had brought the world algebra, geometry, even calculus, and, through it all, negative numbers had never been acceptable solutions. Pascal considered the very idea of zero minus four "pure nonsense," because its result was negative.

Many instances of subtraction—like those tested on the

children at Harvard—don't result in negative numbers. And Pascal was pontificating on math, which is just one abstract representation of our instinct for relative quantity. Still, if we refuse to even conceive of some instances of less, we're certainly not going to pursue them.

Think back to how you must have learned about subtracting, the arithmetic version. Or if you are fortunate enough to know a child who is learning, pay attention to their process. Tangible things, like tempting candies or healthy apples, are often used to help kids understand that abstract numbers correspond to relatable objects. This approach, having a certain number of things and then taking some of them away, induces what math teachers call a "set" schema.

The set schema helps kids visualize less and more—but only so long as the outcomes stay positive. Imagine your six-year-old self presented with the word problem:

Rose has five candies.
Twenty are taken away.

You've just been lied to. Or, as Ben's son quipped when this experiment was run (and replicated) on him, "that is unpossible."

As shown below, our instinct for relative quantity makes it harder to accurately imagine the outcomes of subtractive changes. And if these changes take away more than what was already there, then they are "unpossible."

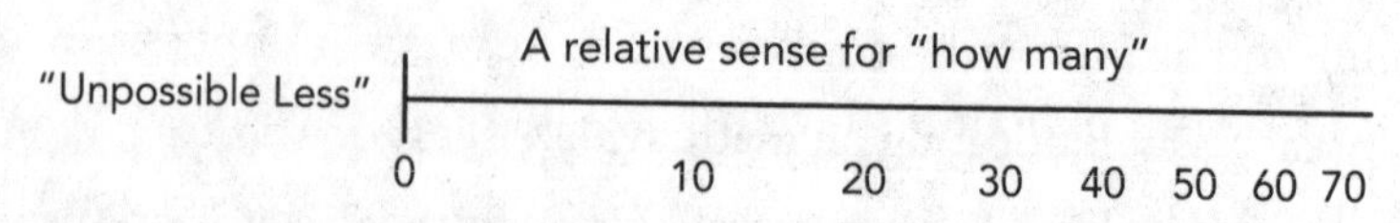

Figure 3: How set schema and our instinct for quantity deter subtraction.

For the set schema, Pascal was right. Negative numbers are nonsense if we use apples or candies (or pea scoops) as the units of less and more.

An alternative to the set schema is the "distance" schema, which, rather than linking abstract numbers to objects, induces us to picture a spatial number line in our minds. Using the distance schema, when seven-year-old you is asked to subtract twenty from five, you will first place each number on your mental number line. Your answer will be fifteen, the distance between twenty and five. Seven-year-old you now has the quantity correct, but not the sign.

Kids begin to get the sign correct when they use the distance schema for a scenario in which negative makes sense to them. A good teacher might, for example, tell her students that the playground is five degrees Celsius during the day and drops by twenty degrees Celsius at night. Assuming the kids have some experience with the metric system (and really cold weather), they can now imagine less than zero. Instead of zero as the lowest possible less, zero becomes the temperature at which water freezes, which reminds us that there are temperatures below that. Subtracting twenty from five results in a very sensible negative fifteen.

The abstract idea of negative numbers defies intuition—until we can map it to some preexisting understanding in our brain. Temperatures below zero work. If your child is a budding actuary, try expenses that exceed incomes. If they golf, try strokes below par.

Shifting from a set to a distance schema is our first example of how changing our view can reveal a route to less. Whether subtracting is an option or "unpossible" depends on the perspective we take.

Let's return to Leo Robinson working to dismantle apartheid. His refusal to unload the *Nedlloyd Kimberley* was a boycott on continued adding to the lifeblood of the racist system. So were the subsequent bans on new investment in South Africa, which is what the United States government eventually did.

If existing investment in South Africa is encoded, whether in minds or in policy, as an unbreachable baseline, then it is pure nonsense to take away beyond that point. It is like a number line that ends at zero.

Divestment required treating that existing investment in South Africa as more akin to the temperature at which water freezes. Sure, the current investment is a notable point on the continuum between not supporting and supporting a racist regime. But the current investment is not an end point. There's a whole world of possibilities on the left side of that number line. Sticking to the set schema in this case is incorrect, and harmful.

It's daunting, I know. To realize our full subtracting potential, we need to override set schema, instincts to add and show competence, and a skewed sense of quantity. So before we set aside math, here's some motivation for our pursuit of less: subtractive changes can be even more powerful than an equal amount of additive ones. What if we were to transform that group of two mammoths by either adding a mammoth or by taking one away? Taking one mammoth away from a group of two results in one mammoth. In this new situation, one mammoth represents 100 percent of the total. Adding one mammoth to the group of two results in three mammoths, and now one mammoth is only 33 percent of the total. Relative to the end state, the exact same change is larger when it brings us to less. No matter how big the mammoth herd, or how big the economy of South Africa, less is the bigger change when we compare it with the new situation.

For better or worse, we will always have a skewed sense of less and more. But math is a nice reminder that we can use our behaviorally modern abilities to override our biological instincts. We can even pick and choose when it makes sense to do so. If Ezra were approximating cookies, I might let him eat just five. But for healthy pea scoops, I can explain that three plus three is, in fact, six.

4.

When I researched the lives of my Neanderthal-mating ancestors, the activities mentioned most were hunting, gathering, eating, and sleeping. With those needs met, my ancestors would pass their time "knapping."

Knapping is the informal but no more intuitive term for lithic reduction: the act of smashing rocks and stones to change them into tools and weapons. It is the task that defined the Stone Age for more than three million years, since we diverged from our predecessor species in the genus *Homo.* So by studying the artifacts of knapping, scholars learn about the evolution of behavior.

One sign my ancestors forty thousand years ago were becoming behaviorally modern is that they had begun to use a new knapping technique. A knapper using the new method would start with a round stone and chip off pieces all the way around the outside edge of the stone. Then, the knapper chipped off pieces from the rounded part of the stone that was facing up. What was left looked something like an upside-down turtle shell, with a rounded surface on the bottom and a flat top surface from which the pieces had been chipped.

Once the stone had been prepared in this way, the knapper would pop off one or more large chunks of rock from this flat and stable surface. Because of the sharp edges from pieces that had been broken off around the circumference, these large chunks were sharp on all sides. Rocks that are sharp on all sides are better for cutting, digging, and hunting.

Before developing this new two-stage approach, knappers would simply chip pieces off rocks and use these pieces. That led to tools and weapons with human-made sharpness on one side and smoother natural contours on the rest. Sharpening these natural contours was impossible because the knappers were then working with a much smaller stone, and risked breaking it, or their fingers. The new method solved this issue because most of the chipping away occurs from the original rock.

The two-stage approach not only made sharper rocks, it also showed behavioral modernity. Chipping away just to prepare the original rock shows abstract thinking, because the knappers had to imagine what the rock would look like after they removed the outer pieces. Michelangelo would later describe his subtractive approach to sculpting: "I saw the angel in the marble and carved until I set him free." The behaviorally modern knappers had, for the first time, seen the sharper knife in the round stone and chipped away to set it free.

I share this because it shows that some of the first evidence of abstract thinking—the very capability that allows modern humans to transform situations—comes from a subtractive change. It's right there in the name: lithic *reduction*. And the two-stage variety was just the latest in a long line of subtractive knapping that defined the Stone Age. My team had found that people add to Legos and Andy's grids, but more than three million years of our ancestors spent their time subtracting from rocks, whether by instinct or choice. So, it seems, not all our inherited behaviors excuse our modern subtraction neglect.

Evolution itself is a marvelous model of balancing adding and subtraction. In finding adaptations that make us more likely to pass down our genes, natural selection does plenty of downsizing. Our modern brains are smaller than Neanderthals', for example, but better for us. Sure, the brain centers for language, social behavior, and decision-making (including to envision change) have gotten bigger. But other parts have gotten smaller.

Evolution also works at the level of ecosystems, and one common result is built-in checks on adding. Whether whales and plankton, foxes and rabbits, or humans and common resources, this balance protects constituents from unbridled growth in one species that might bring down the ecosystem—and the offending species with it.

Closer to home, whereas I crowd my shared research folder with needless files and subfolders, our brains have evolved a built-in protection against overloading our mental processing. When we sleep, our brain cells shrink, which leaves space for microglial cells to come in and clean up unused connections between neurons.

Synaptic pruning is the name neuroscientists have given this automated subtracting. Just as fruit trees grow limbs, we grow synaptic connections between the neurons in our brains. Trees that are watered get bigger and stronger, and the more we use our synaptic connections, the bigger and stronger they get. Of course, a thriving fruit tree also requires pruning, so that precious sunlight or water is not wasted on a branch that will not bear fruit. In our brains, the microglial cells are the pruners. They get rid of less useful synaptic connections so that we can devote more energy and space to the other ones.

To make better use of less, we can gain inspiration from nature. At ecosystem, species, and cellular scales, natural selection works with both hands. We may have instincts to add. But we're surrounded by life that has been transformed by both adding and by subtracting.

As we shift our focus from biological adding forces to cultural ones, let's consider a day in the life of my ancestors. They would wake up, get dressed (or stay dressed—it was an ice age, after all), and maybe cook some breakfast. Then they moved in their bands of about twenty-five people, probably toward rumors of mammoths. By nighttime, they'd be at a new location, where they would set up temporary shelter and knap stones around the fire. Sleep. Hunt-gather. Sleep. Repeat.

Life was a high-stakes camping trip you were born into and never went home from. To survive was to be mobile. There were no home additions to build, obsolete newspapers to hoard, or bags of chips to binge on. Our ancestors had a built-in check on physical adding. When you need to carry all your stuff everywhere you go, you tend to have less of it.

After humans became behaviorally modern, we still spent most of our time following the food supply. Even if you descend from Cleopatra or King Tut, only the last three hundred or so of your roughly ten thousand generations of human predecessors lived in a world with anything close to the civilizations we're now challenged to subtract from. Before then, Leo Robinson would have found no political systems, racist or otherwise. Sue Bierman would have found no obstructive freeways.

Biology does not excuse our modern subtraction neglect. We need to respect the evolutionary inertia that stymies less. Our instincts to eat, to show competence, and even our innate sense for relative quantity can pull us toward more. But whereas evolution relies on random mutations as its mode of change; we humans can be more intentional. Male bowerbirds can't help themselves; I can delete files.

With that in mind, let's move on from my Neanderthal genes to the more recent advent of civilizations, which, as we're about to see, both arose from and are defined by the pursuit of more.

3

The Temple and the City

Adding Brings Civilization, and Civilization Brings More

1.

By about five thousand years ago, north of the Persian Gulf in the fertile lands nurtured by the Tigris and Euphrates Rivers, hunter-gatherer was no longer the sole occupation. Mesopotamian civilization had home builders and architects, teachers and priests, doctors and politicians. These new professionals ate with their families at tables, sitting on chairs, and using utensils. On the menu were fruits and vegetables, pork and eggs, and beer. After dinner, kids could play with rattles, dolls, and balls. And bedtime meant linen sheets and soft wool mattresses, sheltered from the elements by brick homes, which were connected by stone roads, and within walking distance of awe-inspiring temples, ziggurats, and palaces.

With these new things (civilization) came new ideas

(culture). A Mesopotamian Leo Robinson would have found trading boats and docks, and he would have encountered social hierarchies: slaves, laborers, professionals, priests, and nobility. The scholar Elinor Ostrom would have found records of prevailing thought, recorded in cuneiform on clay tablets, and ready for her to rewrite with research.

Like the dawn of behavioral modernity, the arrival of civilization was a major threshold in the history of our adding. Behavioral modernity had given humans the ability to think as we do now, to intentionally change situations, to choose adding and subtracting. With civilization and culture, people gained many of our modern opportunities to add.

In the same way, all these new things and ideas brought new chances to subtract. The painter Pablo Picasso defined art as the "elimination of the unnecessary." *The Little Prince*'s author, Antoine de Saint-Exupéry, observed: "Perfection is achieved, not when there is nothing more to add, but when there is nothing left to take away." If there has been nothing added in the first place, though, there could be no artistic elimination and no perfection-achieving taking away.

Our cultures and civilizations are like our biology in that they change over time, shaped by our surroundings and by our ancestors. But whereas useful genes require generations to spread across a population, useful culture can spread as fast as ideas are shared. Our culture has had far less time to evolve than our biology, but it does so much faster.

We have uncovered biological instincts to add. Now we need

to understand how culture might contribute to our subtraction neglect.

Fortunately for us, civilization not only brought us all these new chances to change the world around us, it also provides a record of what we did. So let's look around.

Rome, AD 2009. The summer sun had yet to rise, and I had the ancient city to myself. As is typical in new-to-me places, I set out on my run to explore, with no particular route in mind. Rome pulled me to the Colosseum.

I had seen plenty of pictures, but none that captured the impression I got in person. Not only is the Colosseum enormous, its details are precise and harmonious, from the spacing of the openings on the façade, to the matching columns on each level, to the etched roman numerals over the gates. The Colosseum would be awe-inspiring even if it were not ancient, the latest stadium downtown or on a college campus. Imagine thousands of years ago, how a new migrant to Rome must have felt when they first encountered the Colosseum.

The rest of Rome was beckoning, but I went no farther than the Colosseum. My awe was undiminished after the tenth lap around the outside perimeter, when I finally headed back to the hotel, where Monica reminded me that my long run had delayed the museum visit she had scheduled for us.

By this time, Monica had learned to live with my attraction to large human-made things. Three years prior, she had spent a day of what I billed as our "Caribbean beach honeymoon" in the middle of Mexico's sweltering Yucatán jungle, because I wanted to see Coba, an ancient Mayan city.

I didn't make Monica run through Coba's ruins with me, but we did climb 120 perilous steps to the top of one of them. Coba's pyramid is more than a thousand years old and tall as a ten-story skyscraper. Mayans built it without motorized construction equipment and also without the wheel. I couldn't fathom the resolve it must have taken to build such a thing in such an era, but at the top of the pyramid, I could appreciate why this would be a good place from which to make sacrifices to the gods. We seemed as close to the sky above as we were to the jungle canopy below.

Rome's Colosseum and Coba's pyramid are records of a surprising kind of adding. These towering displays of excess are so common they have earned an academic name: "monumental architecture." Here is renowned archaeologist Bruce Trigger describing what counts as monumental architecture: "Its principal defining feature is that its scale and elaboration exceed the requirements of any practical functions." The form is literally defined by the fact that it adds well beyond what is necessary.

Monumental architecture is a reliable sign of civilization. Whether the ziggurats of Mesopotamia or the pyramids in Egypt and China, these massive but marginally useful structures grew at the same time as the cities around them.

The cultural steps that led to monuments would have been similar in Rome and in Coba. Romans first achieved surplus food, which led to social cohesion, which brought more dense living, which gave rise to housing, roads, and aqueducts. Then the thriving culture, desiring a place to host staged battles, built the Colosseum. About a thousand years later, and an ocean away, the Mayans went through the same basic cultural progression on their way to adding that ceremonial pyramid in Coba. In each case, behavioral modernity led to agriculture, which led to culture, and then, if a culture really flourished, it might build some big objects that do not provide any shelter. With those steps in mind, monumental architecture seems like the coming-of-age of adding behavior in different cultures.

Except those steps are not how many monuments came to be. In fact, monuments were often added *before* cultures flourished.

There's a more recent example that we can use to see how this works, and to deepen our appreciation for adding as a cultural force. In the late 1830s, the United States of America was not yet a century old and was headed for a civil war that put that milestone in doubt. Yet in the late 1830s, people raised the equivalent of a million dollars to pay for a design competition. They were soliciting ideas for a monument to George Washington in the District of Columbia.

At the time, the entire area was home to about thirty thousand people. The city of Washington wasn't among the ten largest in the United States, which was home to less than 2 percent of the world's population. A monument that cost a million dollars just to conceive, never mind build, was an audacious undertaking for a small group of people with more pressing concerns. Plus, if one wanted to be inspired by a monument to Washington, there was already a perfectly good one about thirty-five miles north in Baltimore.

But the design competition went ahead. So did construction, moving forward despite inconsistent funding, changing political priorities, and the Civil War. After nearly fifty years, the soaring obelisk had reached its full height. When the monument opened to the public in 1886, the United States had nearly quadrupled its population and was now home to the tallest human-built structure in the world.

It's not only that cultures thrive and then decide to build monuments, as was the case with the Colosseum. Monuments are also part of the genesis of cultures. Just as physical activity strengthens our minds, attending to the body of civilization enhances culture. That is to say, monumental architecture isn't actually useless. Sure, Coba's pyramid and Washington's monument were a lot of adding for no shelter in return. But these projects brought large groups of people together, first to plan and build them, and then to be awed by them. And when large groups of people come together, that's when you get culture and civilization.

Perhaps you remain skeptical that culture arose from monuments to *more*. Even as a lover of big structures, I understand the hesitation. Does monumental architecture really deserve a place alongside government, religion, and writing? Does it really have to be there for a group of people to be considered a civilization? But those who study ancient civilizations have moved on. The question is not whether monumental architecture should take a back seat to these social changes. The question is whether it should be elevated above them.

Modern analysis suggests that monumental architecture may have been the catalyst for civilization. Göbekli Tepe, which translates as Potbelly Hill, is an archeological site located in what is now Turkey. In the 1960s, a team of researchers surveying ancient sites in the region had visited, seen some small broken slabs of limestone, and moved on. In their written report, they concluded that Potbelly Hill was probably an abandoned medieval cemetery.

A few decades later, the archaeologist Klaus Schmidt read the researchers' original report. Schmidt wasn't sold on their conclusion that the mound, which is about fifty feet high and a quarter mile across, was just a cemetery. So, he paid a visit to Potbelly Hill.

It didn't take long to find the first big rocks. They were so close to the surface that they had been scraped by farmers' plows. When Schmidt and his team dug deeper, they found massive stone pillars arranged in circles. At the center of each circle were even bigger pillars—about as tall as giraffes, and more than ten times as heavy. Remarkably, the ancient civilization had somehow carved, moved, and stood up these gigantic pillars.

But it's what Schmidt and his team didn't find on Potbelly Hill that really made archaeologists and historians think. There were no cooking hearths, no houses, and certainly no toy rattles. There were none of the telltale signs of a permanent settlement that were all over nearby sites of about the same age.

What Schmidt found and didn't find at Potbelly Hill blew a hole in the prevailing theory that people began farming and staying in one place, and only then acquired the time, skills, and resources to add monuments and everything else that goes into civilization. Hundreds of people, working together, would have been needed to extract, move, and place the temple's stone columns. Yet the temple predates villages and even agriculture in the region. In a world where hunter-gatherers had been hanging out in packs of about twenty-five, it would have taken unprecedented cooperation between groups to build the temple at Potbelly Hill.

Schmidt laid out a new-and-improved theory in his article "First Came the Temple, Then the City." As that title asserts, and is increasingly believed, building the temple at Potbelly Hill could have been what brought the hunter-gatherers together in the first place. The idea of the temple provided the first excuse for disparate bands to convene. Then, the long-term commitment needed to build and then maintain it drove the hunter-gatherers to seek less transient food sources. This, Schmidt contends, is what led to agriculture. Findings from sites around Potbelly Hill confirm that, within the thousand years following the temple's construction, settlers had domesticated wheat and corralled livestock.

The findings at Potbelly Hill reverse the relationship between civilization and adding. Scholars had long believed that no matter how eager a group was to begin adding temples, they had to first learn to farm and live in settled communities. But Schmidt said it was the other way around; the long-term effort to build the temple was the impetus for farming. In this way, adding brought civilization.

A cultural tendency to build would be enough to help explain why we neglect subtraction. But as civilizations appeared, so did another time-honored kind of more: our material culture. Apparently, the fourteen different styles of sneaker in my closet are an extension of practical variety that let people navigate their new social lives.

Material culture helps us live together in big groups, even though our brains evolved in far smaller ones. In a band of hunter-gatherers, everyone could learn one another's traits, skills, and favored cuts of mammoth meat. As civilizations grew, this personalized approach became impossible. But humans still needed to make sense of the people around them. Material culture responded to that need.

Instead of trying to keep track of thousands of individuals, humans could lump their neighbors into a more manageable number of categories, based on their clothes, beads, and so on. These physical things made interactions with strangers more predictable, as when you walk into a deli and recognize that the attentive person with the apron and pad of tickets will take your order when you are ready. Or, inversely, how I can walk across campus undercover—no longer a professor—just by donning shorts, a T-shirt, and one of my (fourteen) pairs of sneakers. In both cases, material culture suggests how people should interact even though they have never met.

As with adding to our skylines, there is no doubt that adding to our closets has roots in the beginning of civilization. Again, the question is "Which came first?": whether material culture may have predated and helped spark civilization. A theory now gaining momentum is that hunter-gatherers randomly generated material culture—say, risky hunters wearing mammoth skins and cautious hunters donning rabbit hides. In this example, no longer would each hunter-gatherer need to memorize the hunting-risk tolerance of everyone else in their band—that could be inferred from dress. Such mental shortcuts based on material culture would have allowed the hunter-gatherers to manage relationships in larger groups. The larger groups, in turn, would have allowed the hunter-gatherers to build monuments and hierarchies, thus growing civilization.

A cultural pull toward monuments and material trappings is counter to subtraction—whether it's Ezra building more Lego towers or me buying him more Legos. All over the world, too, civilizations spurred another engine of adding: writing. People poured their newfound time and energy into this new medium. Writers could keep records of who owed who, which brought more trading. Writers could create transparent and consistent laws, which enabled bigger civilizations. Writers could convey more ideas and convert them into more things.

Writing both showed adding and enabled it, releasing the

capacity to accumulate information from the confines of individual minds. Fleeting ideas could now be made to last, extending the time across which one person could share accurately with another. Klaus Schmidt's work at Potbelly Hill could build directly from the thoughts of researchers he never met. Ideas of all kinds could spread all over the world soon after someone like Stephanie Preston, Elizabeth Spelke, or Elinor Ostrom discovered them.

Archaeologists will continue unearthing the origins of civilization, but our science of less doesn't need to wait. There is no question that adding and culture are inseparable. In key ways, as we've seen, the earliest civilizations were defined by more. Humans who no longer needed to spend all day searching for food added more things: pyramids, buildings, and clothes. They added social structures and ideas too: laws, religions, writing, and math. For people living in a world lacking all these things and ideas, it would have seemed unnatural to subtract them. It would not have been that there was "nothing *left* to take away," but rather that there was nothing to take away, period.

Sure, cultural evolution subtracted some things. There was less hunting and gathering, but civilization was a project of enlargement, and since modern culture arose from these first civilizations, we all share the heritage of adding at a "scale and elaboration which exceeds the requirements of any practical functions."

2.

Our shared love of more may be strong, but it is not monolithic. Over time, cultures evolve to produce distinct views of the world, and these views shape how we think about change. As we investigate the cultural forces behind our subtraction neglect, then, we should see what diverse worldviews might have to say about less.

For that we need some foundational knowledge from Stanford professor Hazel Rose Markus, who pioneered the study of how our mental habits relate to our cultures. Markus's influential research is the basis of *Clash!*, her coauthored and often firsthand account of research on cultural worldviews. In Markus's book, the "clash" is between those who define themselves by their own abilities, values, and attitudes (independent), and those who define themselves by their relationships, social roles, and group affiliations (interdependent).

Two traditional origin stories—yin and yang, and Cain and Abel—illustrate how cultures can perpetuate these "interdependent" or "independent" views of how the world works. Each story has been used to explain how humans came to be.

Yin and yang were born from chaos when the universe was first created. The pair achieved balance in the cosmic egg, which allowed for the birth of the first human and the first gods. To this day, yin and yang exist in harmony at the center of the earth.

The brothers Cain and Abel were the first sons of Adam and Eve. Cain became a farmer and Abel a roaming shepherd.

Each brother gave some of their food to God. God liked Abel's gift better than Cain's. Jealous, Cain killed Abel. God sentenced Cain to a life of wandering. Eventually, he and his descendants built the first city.

Yin and yang reflect the interdependent view, in which people see themselves as adjusting to the rest of the world, as balanced complements, just like yin and yang in the cosmic egg. The Cain and Abel story implies that humans are independent from situational constraints. One can be a shepherd or a city-builder.

Certainly, as Markus makes clear, there is more variety in cultural worldviews than any origin story can capture. Chinese culture leans more interdependent than United States culture, for example, but Chinese Taoism, with its emphasis on individuals' happiness, trends more toward independence than Chinese Confucian culture. In the independent United States, people in small Southern or midwestern towns who prize family and community relationships tend to see themselves as more interdependent than people in the make-it-here cities on the East and West Coasts.

The culture clash is not limited to religions or addresses. There are general variations in worldview between people in developed and developing nations, between men and women, and between blue- and white-collar occupations. Full disclosure: I check every box for independence, for the Cain and Abel view of the world. The global north tends to be more independent than the global south, as do men, working professionals, and those who aren't religious. In the beginning, my parents named me Leidy. There's even research showing that people with unique names are more independent.

Perhaps, like me, you hold out hope that we can chart our own wandering path. Or maybe you see destiny as returning to balance in the cosmic egg. In any case, let's see how the contrasts between them might influence our adding and subtracting.

Since before Hazel Rose Markus made cultural psychology a field, researchers have suspected that our worldviews influence how we view the situations we encounter (and try to change). In the late 1940s, the psychologist Herman Witkin sought to test this by changing his subjects' perspectives—literally. In his laboratory at Brooklyn College, Witkin and his fellow researchers built a big box, large enough to comfortably fit a person, framed with two-by-four lumber, and tiltable, thanks to a motorized jack attached to the base of the box. This tiltable room allowed the researchers to measure what Witkin would eventually call "field dependence." Witkin took the term *field,* which we will continue to use, from the work of Kurt Lewin, another eminent psychologist whose revolutionary "field theory" had drawn attention to the need to understand behavior as a product of people *and* their surroundings—their field.

In Witkin's tiltable room, the people could be moved, and so could the surroundings. Participants were seated on a chair, which was then tilted at various angles. The room was tilted too—unbeknownst to the participant. The participants' field dependence was the extent to which their view of their own tilt depended on the tilt of the room. If both the chair and the room were tilted at the same angle—say, fifteen degrees—a participant who reported being tilted at nine degrees would be less field dependent than one who reported being tilted at three degrees. The participant who perceived nine degrees of tilt was influenced less by their relationship to the room, or field. The participant who, in this same situation, perceived only three degrees of tilt was more dependent on their relationship to the field. The tiltable room studies provided some of the first evidence of systematic variation in how different people view the same situation.

Measures of field dependence eventually got more practical. Embedded Figures Tests measure it just as well and can be done on paper or a screen. Participants are given schematics like the one shown below.

Can you find the octagon?

The longer it takes you to see the octagon in its surroundings, the more field dependent you are. If you are paying attention to all the lines in the schematic, the field, it takes longer to focus in on the octagon. Alternatively, if you find the octagon quickly, you might not take in everything that is around it.

One inference here is that independent types might overlook subtraction when, in our focus on individual objects, we fail to consider taking away from the surroundings. Advice to counteract this general oversight goes by many names. Undergraduate design students are taught to consider both the figure and the

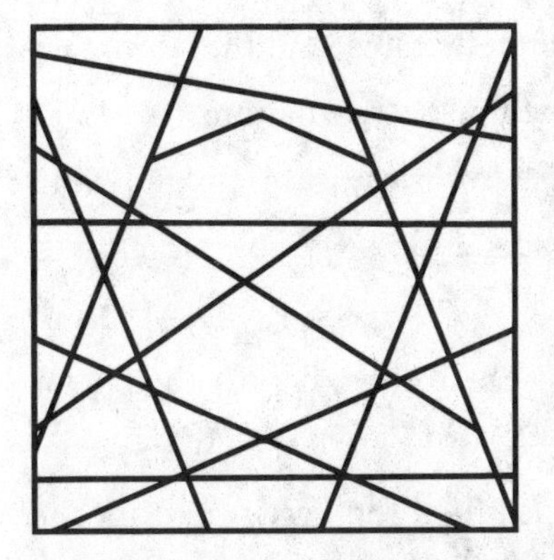
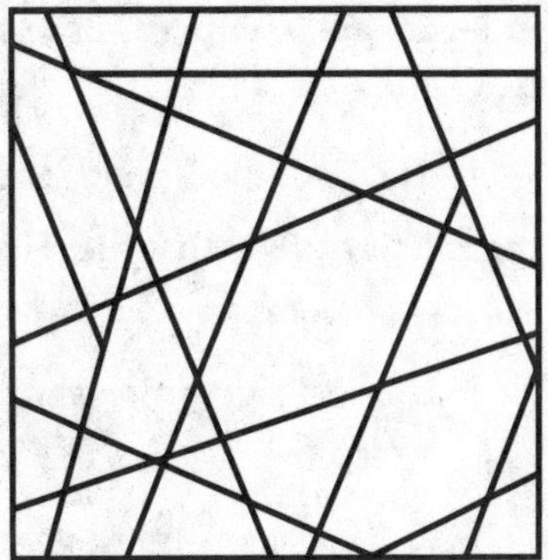

Figure 4: An Embedded Figures Test

ground. By the time they graduate, they will have heard professors wax on about positive and negative spaces, explicit and implicit ones, solids and voids.

Of course, another valid inference is that interdependent types might overlook subtraction when, with attention to the surroundings, they overlook objects in need of removal.

Those speculations aside, the main lesson here is that our view of the field has roots in cultural evolution. To see how, imagine yourself in this study: first, you are shown a pile of sand, which has been formed in the shape of the letter *S*.

Then, you are shown two more situations: 1) a pile of sand, and 2) glass pieces that have been formed in the shape of an *S*.

You are then asked, "Which of these second two situations is most similar to the first situation you saw—the sand shaped like an *S*?"

Mutsumi Imai and Dedre Gentner have given versions of this ingenious test to thousands of people. Imai is a professor in Japan, and Gentner in the United States, and they compare responses between people in their respective nations.

In the S-shaped sand study, American participants were more likely to say that the *S* made of glass was more similar to the *S* made of sand.

Japanese participants were more likely to choose the pile of sand.

The Americans mentally classified the first situation, the S-shaped sand, as an *S* that just happened to be made of sand. The Japanese coded the very same situation as sand, which happened to be shaped like an *S*.

It's tempting to presume from these findings that Japanese people are more holistic thinkers, or that people in the United States are better at seeing details. A more useful and defensible conclusion is simply that some people see the field (the sand), whereas others see objects (the *S*).

Different people can see the exact same situations in predictably different ways. That's why field dependence matters to us. Because how we view a situation shapes how we try to change it, whether we think of, and then choose, subtraction.

The rewards of seeing the field are on permanent display in Savannah, Georgia. When Monica and I visited, we took two extra days to wander the historic coastal city. The aerial view of Savannah looks a bit like one of Andy's grid tasks. As shown

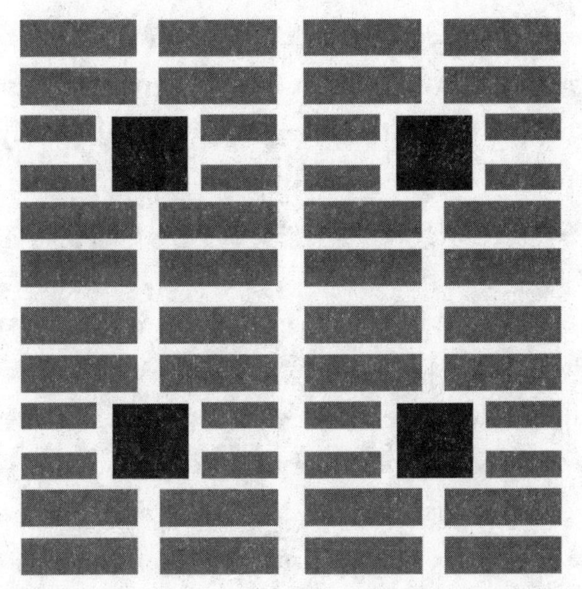

Figure 5: Four wards, and four squares, of Savannah's plan.

below, the city is made up of matching wards. At the center of each ward is a square. Each square is surrounded by eight city blocks: four smaller commercial blocks adjacent to the square and bordered by wider streets, and four larger residential blocks in the corners each split by a narrow lane. It's a simple design that has endured to this day. And it's a planning masterpiece.

I was surprised to learn that Savannah's design was enlightened in more ways than one. James Oglethorpe, who founded Georgia and drew up the plan for Savannah, was very progressive. Georgia was founded to house relocated inmates—debtors released thanks to Oglethorpe's push for prison reform in

England—and under his watch, which lasted until 1742, slavery was prohibited in Georgia. So, for that matter, was aristocracy. Oglethorpe's vision of social equity extended to his layout for Savannah. By serving as shared public space, the now-famous squares fostered community engagement and participation in city affairs.

I suspect any urban planning graduate—in the United States, at least—could sketch a close approximation of Savannah's layout from memory. But one need not have a design degree to appreciate Savannah's straightforward plan. Monica and I, like so many others, were lured through the historic city, sensing that the next square was just around the corner and eager to see what would be unique about its trees and flowers, paths and benches, bricks and iron. At the same time, the city was teeming with people going about their daily lives in offices, homes, and restaurants. One was never far from a new social interaction, in a square or on a street—or from social refuge, on a secluded bench or a side alley.

Savannah's plan works for visitors and residents. And the plan works by seeing the field; by considering the field of public space, the city reduced the size of private building lots. Most cities mandate far roomier building lots than Savannah has, and with good intent. Open space not only looks nice, it also improves air and water quality. But in Savannah, the public squares, and the streets and alleys connecting them, function as open space. Everyone's lot can be smaller since everyone is close to a park. Savannah is unforgettable because the public squares and private lots were considered together.

Whether planning a colony or solving Andy's grids, we're less apt to neglect subtraction when we see the field: all those things that may recede into the background, but nevertheless determine our options for change. The field is not limited to physical things. For Oglethorpe, the design of Savannah was part of a larger idea of social equity. He was designing for people and their relationships. By seeing this less visible part of the field, Oglethorpe created the captivating streets and squares.

Believe it or not, the story of Harlem's hoarding Collyer brothers has something to teach us here too. The brothers' demise is a reminder of our adding instincts. What happened next represents how hard it can be to see the field, and how powerful it is when we do.

After the brothers died in 1947, their brownstone was demolished, and the lot on Fifth Avenue and 128th Street sold at auction. The lot sat as an eyesore until the early 1960s when, in part because of riots in Harlem, national attention was drawn, as it periodically is, to the plight of people in low-income urban communities. One suggested response was to turn vacant lots into "pocket" parks.

To try out this idea in Harlem, philanthropists bought three separate lots on 128th Street. One lot became a playground for little kids; one became a space for basketball and Ping-Pong; and one—the Collyer lot on the corner of Fifth Avenue—was made into a sitting area for grown-ups.

The philanthropists knew that opening these spaces to the public would not address every systematic disadvantage faced by people in the neighborhood. The philanthropists also knew, however, that one pervasive hardship in low-income urban areas is a lack of outdoor space.

Pocket parks were a shift in thinking about outdoor space in cities. Whether Savannah's frequent squares or New York's vast Central Park, people were used to spaces that had been planned and designed in much the same way as the buildings that grew up around them, taking a blank slate and changing it by adding paths, ponds, and monuments.

Pocket parks do not lend themselves to such an approach. Collyer Brothers Park on Fifth Avenue and 128th Street depends more on the surrounding field than on the vacant lot itself. The sense of shelter and seclusion in the park's sitting area is provided by the adjacent building on Fifth Avenue. The few trees and bushes in the pocket park are strategically placed to provide a see-through buffer between the sitting area and the bustle of the sidewalks and streets. Collyer Park depends on its field, which is why the American Society of Planning Officials noted in their original report on pocket parks that they require "a minimum of expense, but a great deal of imagination." That insight predates—but sounds to me like—my team's finding from the storm trooper experiment. Subtracting the single Lego may require less effort. It may even cost less. But to transform via subtracting requires more thinking.

The good news is that, whether in Legos or in Harlem, the extra thinking needed to arrive at less can pay off. The Washington Monument reflects more power and Coba's pyramid is a more convincing path to the gods because these things are surrounded by open space. And because it contrasts with the surrounding city, Collyer Brothers Park stands out in one of the hardest places in the world to do so.

What is true in things is mirrored in the world of ideas. As theories are laid down, taking away from this conceptual field can enhance our understanding. It was true for Elinor Ostrom, and it was true for Klaus Schmidt. Surely there was some initial confusion and maybe even frustration as his team, having immediately uncovered the awe-inspiring monuments, dug and dug, finding no signs of permanent settlement on Potbelly Hill. The prevailing theory couldn't explain monuments without settlement. But, by paying attention to what wasn't there, Schmidt came to a more accurate diagnosis of cultural evolution.

Culture reinforces our adding tendencies, yes. But it is also civilization that presents us with all the situations we can improve by subtracting. Our crowded cities and knowledge have given us a field in which pocket parks and subtracted ideas are not just possible but transformational—if we can stop neglecting them.

3.

On the Mall in Washington, D.C., I am free to race past the Washington Monument but am reminded by signs and workers not to run through the memorials. Reminders are redundant for the Vietnam Veterans Memorial. The Wall is so awe-inspiring that, even though there is a welcoming and gradual path heading right in front of it, I find it impossible to run by without stopping. Whereas the Washington Monument was once the tallest structure in the world, the Vietnam Veterans Memorial is recessed into the ground, a series of plain black granite slabs. If you haven't been, I think Maya Lin's original description of her design perfectly captures the first impression:

> Walking through this park-like area, the memorial appears as a rift in the earth, a long, polished, black stone wall, emerging from and receding into the earth. Approaching the memorial, the ground slopes gently downward and the low walls emerging on either side, growing out of the earth, extend and converge at a point below and ahead. Walking into this grassy site contained by the walls of the memorial we can barely make out the carved names upon the memorial's walls. These names, seemingly infinite in number, convey the sense of overwhelming numbers, while unifying these individuals into a whole.

On the D.C. Mall, the Wall is different. Maya Lin subtracted.

Determining design intent can be maddingly subjective; one might protest that Lin added the wall to the earth, that she only took away soil in service of a grand vision of adding. Thankfully, we have Lin's own words to interpret how she approached her masterpiece: "I saw the Vietnam Veterans

Memorial not as an object placed into the earth but as a cut in the earth."

That settles that. Lin subtracted. And in doing so, she inspired awe in the arena of monuments.

Lin famously conceived her iconic design when she was still an undergraduate at Yale University. An army colonel visited her dorm room to tell her that she had been selected out of more than 1,400 entries in the competition to design a memorial for what, at the time, was America's longest war. The choice has proven to be a good one; the Wall is among the most cherished monumental architecture in the United States, often

Figure 6: Maya Lin's portrayal of her winning design for the Vietnam Veterans Memorial

beating out the Washington Monument on subjective lists of such things.

What is now treasured was once controversial. When Lin's design was unveiled, politicians withdrew their support for the memorial, at least in part because it was not monumental enough. Public outcry was enough to give the secretary of the interior an excuse to delay issuing a building permit. The Wall prevailed, but not without a compromise that added a flagpole as tall as a five-story building, and a bronze sculpture of larger-than-life soldiers off to the side of the recessed wall.

While Lin's design was undoubtedly shaped by who she was, she has questioned whether it would have been chosen had the jurors known her background. Lin's design was submitted anonymously; had it not been, her young age could have been an impediment, as could her gender. Architecture is so male-dominated that one of the comments made by a juror about Lin's painterly soft pastel drawings was: "He must really know what he is doing to dare to do something so naive." After Lin was revealed as the designer, the reflective black granite on the Wall was criticized as "too feminine." If so, then good. This polished stone is precisely what creates the haunting reflective space that appears to be on the other side of the names, what Lin called the "space we cannot enter and from which the names separate us." Evaluation of Lin's design could also have been biased by her ethnicity. After Lin was revealed as the designer, *The Washington Post* published an article referring to "An Asian Memorial for an Asian War." In her very first press conference,

Lin was asked why someone of Asian descent should design the memorial for the Vietnam War.

It is tempting to look for links between cultural background and approaches to design, subtraction included. Whenever people learned of our research, speculation about cultures that would be better or worse at finding less wasn't far behind. Gabe's friend, who is from the Netherlands (and was drinking a beer at the time), was certain that people like him, from Germanic cultures, would all make the right subtractive change on Andy's grid task.

Such speculation begged for evidence.

Gabe called in favors from her international collaborators. A contact in Japan agreed to test Andy's grids with students in that more interdependent culture. Through a contact in Germany, we would test whether Gabe's friend's confidence was warranted.

It was not. We found that students in both Germany and Japan neglected subtraction too. There was some insignificant variability between the different groups, but this was smaller than the variability between different grid patterns.

My team's findings from Germany and Japan don't mean that there are no cultural or geographic differences in the use of subtraction. It's just two nations and limited participant pools. Other cultural differences between the groups we tested may also have canceled out geographic ones. Maybe people who have less see more value in adding. Perhaps people who live in densely populated areas have learned to subtract. It is impractical to try to isolate all the possible cultural variables that might make a difference.

Speculating about cultural differences is fun to do over drinks. But we can't isolate most behaviors (adding and subtracting included) to independent or interdependent, women or men, undergraduate or "grown-up." Maya Lin's parents had migrated to the United States from China—but we need to resist any urge to explain Lin's monumental subtraction as a product of Eastern, interdependent culture. For one, Lin recounts not even realizing she was ethnically Chinese until later in life.

All of my team's experimental evidence suggests that subtraction neglect is robust across groups and situations. More importantly, even if we were to find differences, there isn't much we can do about it. "You should have been born in the Netherlands" is not an actionable recommendation. Even if it were, it wouldn't make you any better at finding less.

That said, what at first seems a practical dead end has useful takeaways. To see how, let's return to Hazel Rose Markus and her study of mental habits related to culture. In *Clash!*, Markus emphasizes that "we all have an overriding sense of our self as the same across all places, times, and situations." That sense is wrong, she says, because, "when we look more closely at the story of our lives, we see that we actually have many different selves within our one self." Markus is a different self when she is surfing from when she is teaching, from when she is writing an academic paper, from when she is writing a popular book. Likewise, I may trend independent, but when I am parenting,

I try, or am forced, to be more interdependent. It's pretty hard to think you control your own destiny when a three-year-old is ordering you to go get him a glass of milk so he doesn't have to stop building Legos.

With her research, Hazel Rose Markus confirms that Walt Whitman had it right with his inner dialogue in "Song of Myself":

Do I contradict myself?
Very well then I contradict myself,
(I am large, I contain multitudes.)

To see and get to less, what matters is not Chinese or American, women or man, or even our interdependent or independent selves. What matters is accessing our "multitudes." As Markus puts it in *Clash,* "For many people, the realization that you have two equally legitimate selves marks a 100 percent increase in psychological resources."

With our newfound appreciation for cultural multitudes, this advice becomes more useful than the generic "think differently." It suggests specific unique and expert views that we can use to increase our chances of seeing subtraction. When I struggle to cut a paragraph that isn't relevant to this book but that shows my independent competence, I can look to one of my interdependent multitudes. Then I see that the paragraph might break trust with precious readers like you, and it's gone. (Well, saved in an "excerpts" file that runs nearly forty pages.)

My multitudes are few compared to Anna Keichline's. Keichline made one of the most ingenious engineering advances of the twentieth century, but she would not have identified as an engineer. By degree and license, Keichline was an architect, one of the very first graduates of Cornell University's top-ranked program, and the first female architect licensed in the state of Pennsylvania. And Keichline brought more than disciplinary multitudes to her designs. She had played basketball in college, was one of the first women to drive a car, and had been a spy during World War I, emphasizing her strength, mechanical ability, and fluency in German as she made her case to be assigned to a "more difficult . . . [or] more dangerous" wartime role. Anna Keichline was large; she contained multitudes.

Today, most building blocks are hollow, often shaped like a sideways and squared-off number eight. But before Anna Keichline, building blocks were solid. Solid blocks built Mesopotamian homes, Rome's Colosseum, Coba's pyramid, and the Washington Monument. If your house is more than a century old, it probably rests on solid blocks. But, in her 1927 patent for the K-brick, Keichline subtracted that mass. Keichline the engineer knew that as long as the load-bearing outside parts of the blocks were solid, the insides could be hollow. And Keichline the architect knew that hollow blocks would appear exactly the same from the outside.

Not only was less possible, less was better. Keichline's hollow block required half the material compared to what was then the typical building block. This made her version less expensive and

Figure 7: Anna Keichline's K-brick (Photo by Nancy Perkins)

easier to build with. The block took less energy to make and less fuel to transport. And because of the insulation provided by the air voids in the individual blocks, the resulting buildings were more comfortable, less noisy, and less prone to fire.

Keichline's subtractive insight led to the building block that is now used to build everything from the hollow brick façades of schools and skyscrapers to the foundation walls for my two-story addition, which rests on a few hundred hollow concrete blocks.

I doubt Keichline would have come up with her improvement to engineering if not for her architectural multitude. Her block is a magnificent example of considering objects and the field together—or, as they would have been called in Keichline's classes at Cornell, the figure and the ground. Keichline subtracted material from the object, the concrete part of her brick. Doing so allowed her to enhance the field, creating the space that provides the light weight and insulation that make her brick better.

Few have access to all the multitudes of Anna Keichline, but we all have access to some. Think of accessing our multitudes like the subtracting views we tried in the last chapter. When the set schema doesn't reveal subtraction, we might try the distance one. Or, when the absolute view doesn't let subtracting add up, we can check the relative one. And, if an independent focus on objects leads to adding, maybe an interdependent appreciation of the surrounding field will show you less.

Admittedly, consulting two multitudes requires double the thinking, which we don't always have the time for. And while some of our multitudes may help us see subtraction, others may show us adding—reinforcing our distorted approach to change. No single self is always helpful on our search for less. But, there is a cultural approach that just might be.

4.

When sharing my research, I have learned to mention that just because people overlook subtraction, I am not contending that it is always the best choice. No matter how many disclaimers I make, I'll inevitably get a question about some freeway, meeting, or idea that should most certainly not be subtracted.

I empathize, because I still catch myself thinking about

adding and subtracting as an either-or proposition. One of my final passes through these pages was to make sure I was representing adding and subtracting as complementary. Let me be clear here: to find the options we are missing, we need to go from thinking add *or* subtract to thinking add *and* subtract.

How we deal with contradictions, whether we think add *or* subtract or add *and* subtract, is a key difference in how people approach change. Which do you tend to do?

Let's practice by picturing Kali.

According to Rolling Stones' lore, the band's famous Hot Lips logo was inspired by the Hindu goddess of destruction. The Stones' logo features a pronounced and cartoonish set of red lips framing a long red tongue, defiantly sticking out at the world. The Hot Lips ooze lust and rebellion, just like the legendary rock and roll band.

In addition to the full lips and defiant tongue, depictions of Kali usually include blood-soaked weapons, a girdle of detached arms, and a necklace of severed heads. Those bloody features are not included in the Stones' logo.

Lots of gods create. Kali takes away. She is the goddess of destruction, and she sounds terrifying. As the story goes, a group of mortals once tried to win Kali's favor by sacrificing another human to her, but she disapproved of their offering. Whereas Cain and Abel's God sentenced the murderer Cain to wandering, Kali was less forgiving. She decapitated her misguided worshippers—and then she drank their blood. Another greatest hit is when she took on a fighting demon who was dripping blood, with each new drop turning into another fighting demon clone. She resolved the situation by draining the original demon's blood, which allowed her to focus on the demon replicants, whom she ate.

Kali's battle against these clones is the origin story for the Thug cult, who considered themselves Kali's children, created from her sweat. As recently as the 1800s, the Thugs rampaged around India, Bengal, and Tibet, killing and robbing travelers. A typical Thug victim, of which there may have been up to two million, was strangled and mutilated in honor of the goddess of destruction.

Don't get rid of your Hot Lips T-shirt just yet. Because Kali is also a symbol of motherly love. She is creative, nurturing, loving, and benevolent.

What?

Kali's duality makes me uncomfortable. I'm usually tuned to a multitude with clear categories of good or evil, win or lose, and, until relatively recently, add or subtract. If one idea contradicts another, then one of the ideas must be wrong. If A is true, then not-A must be false.

Kali is one big contradiction. She smiles at us, even as she kills. The severed heads and arms that adorn her body show that she has acted on her rage, and they also symbolize creative power and severance from the human ego. Her famous tongue signals lust, rebellion, and a thirst for blood. According to those who study such things, it also signals modesty and shame.

Instead of fitting into a comfortable narrative of good versus evil, Kali transcends both. She is Cain and Abel—and God—all wrapped into one.

Resolving contradiction is not a bad thing. Doing so has aided reasoning at least since Aristotle, who held that if one idea contradicted another, then one of the ideas had to be rejected. We have this logical reasoning to thank for all sorts of scientific breakthroughs, everything from our single, repeatable, biological classification system to the mathematical logic that led to modern computers, to Elinor Ostrom subtracting the tragedy of the commons.

The problem comes when we try to resolve contradiction between ideas that aren't actually in conflict. As we've seen, for example, the question is not whether biological or cultural forces explain our adding. Both play an overlapping role in our failure to subtract. Arguing about which is the true culprit only wastes time and distracts us from learning how we might do better.

Trying to resolve false contradiction is one reason, despite Hazel Rose Markus's best efforts, we pigeonhole people and cultures into rigid categories. Trying to resolve contradiction is what causes us to argue whether the billionaire who donates a new building to her alma mater is doing so to avoid taxes, to erect a monument to herself, or because she cares about education. It's hard to accept that it's probably a bit of each.

That's why I'm constantly reminding audiences that I am not necessarily in favor of removing the local freeway. The question is not, "Should we add or subtract?" The question is, "How do we use both?"

As we saw in the last chapter, our biology both generates *and* selects. In our culture, the fact that adding has brought us so many great things can only mean that neglected subtracting has latent potential. Thinking add *and* subtract even works for monuments; while Lin may have seen the wall itself as a subtractive "cut," she also added in a way that few of her competitors did. Lin's memorial includes all 58,318 names of the people who died fighting for the United States in the Vietnam War.

Cultures born from adding keep adding, which means they crave more. More food, more shelter, and more infrastructure. Professional governments and militaries are eventually needed, which, in turn, require new roads, forts, and defensive walls. These reinforcing needs demand more natural and human resources, and to get them, adding cultures have expanded. Romans built the Colosseum using treasure they took after killing around a million people in Jerusalem, more than a thousand miles east of where the Colosseum now sits.

When an expanding Roman Empire annexed a farming settlement near Jerusalem, the resulting culture eventually prioritized Roman-style adding. It was no different for the spread of the Mayans throughout the Yucatán Peninsula. When the adding that built Savannah and the Washington Monument spread westward across the continental United States, it subsumed the Native American way of life. These adding cultures eventually turned into us.

All that said: even as adding built civilization, plenty of people remained suspicious of, or foreclosed to, more. Warnings against too much are a common theme across all the major religious texts. For some sects—Franciscans and Calvinists, Zen Buddhists and Hindu ascetics—spirituality meant active disdain for worldly accumulation. And for most people in premodern independent and interdependent cultures, for a Roman soldier and for a Mayan builder, the only plausible socioeconomic goal was to maintain one's station in life, not advance it.

Being unable or unwilling to pursue more is not the same as seeing the value in less, but it does help keep some adding in check. So even as adding grew culture, the quest for more as we know it had not diffused throughout society.

Yet.

4

More-ality

Time, Money, and the Modern Gospel of Adding

1.

When I told Monica I was working on a book about why we neglect subtraction and how we can do better, the first thing she asked was, "What if readers find out about our addition?" Let's see.

Monica was referring to how we transformed our new (circa 1947) Cape Cod–style home. At 1,500 square feet, our new home was about half as big as the home from which we had moved. We didn't mind downsizing, but a renovation was in order. Our new home had been a student rental, and while the tenants had treated it well, the landlord had made questionable decisions. My nemesis was the vinyl floor in the kitchen, which was white-black checkered and also textured, presumably to better hold and display dirt.

To gather ideas for our renovation, I ran a design contest for students. Monica and I had clear intentions. We advertised the contest with the title "Addition by Subtraction" and emphasized that our goal was "to improve the human experience in the home and surroundings by subtracting." We were willing to pay more, we said, if our renovation could make a statement through subtraction. Conscious of exploiting students' expertise, we offered $1,000 in cash prizes, plus free cookies. A couple of dozen architecture, engineering, and environmental design majors signed up.

I inspired the entrants with less-is-more design wisdom I had curated over the years: industrial designer Dieter Rams's advice: "Less, but better"; a TED talk by the peerless chemist George Whitesides, in which he says we need to respect simplicity, and stop dismissing it with the same "we know it when we see it" logic we use for pornography; and even a bit by the late comedian George Carlin, in which he shares his philosophy on house design: "If you didn't have so much stuff, you wouldn't need a house . . . A house is just a pile of stuff with a cover on it." At the culmination of the three-month competition, entrants presented their proposals to a general contractor, Monica, me, and Ezra, who was partial to those who regifted cookies.

The students came up with clever designs. A sophomore found unused vertical space and used it to add a lofted area in Ezra's bedroom. A junior changed the grading of the backyard to provide outside access to the basement, turning it into a viable living space. A graduate student pair reconfigured the entire interior floor plan.

And yet—none had actually subtracted. Our competition theme had inspired economy of space, but not less of it.

As Monica worried you would learn, the professor was even worse than the students at taking away. A five-room, two-story, nine-hundred-square-foot addition now extends from the rear of what had been a little Cape Cod.

Sure, adding improved our home: we have nice new space, the first story of which serves as a cover for Ezra's Legos. But we never got to see if our home would be improved by subtraction as well.

There are interwoven psychological, biological, and cultural forces behind why we neglect subtraction, as we've seen. But none could fully explain my addition. The design competition entrants and I had certainly thought of subtraction. It was the theme of my whole contest. Yet I, a behaviorally modern human, suppressing my instinct to add and resisting outside pressures, failed to subtract.

To explain this choice, we have to consider one last force against less: the economy. For our home renovation, the sticking point was the fact that a home's value increases with total square footage. Entrants in our design competition could not figure out a way past this financial reality, and neither could I. Spending money without adding square footage would have been a risky investment. Spending money to get rid of existing square footage was preposterous. Less space was as absurd as negative numbers were for Ben's son and for Blaise Pascal. Someone else could be the home-subtraction pioneer.

This idea of adding square footage to grow personal wealth is relatively new. While humans have always had instincts to add, and our ancestors' cultures of more turned into us, modern economies overlay their own logic on how we build, how we think, and how we spend our precious time. So just as we have probed biological and cultural roots of our adding, we need to probe the economic forces that contribute to its overuse. And to appreciate this relatively new pull toward more, we need to understand its origins.

On January 20, 1949, more than a million people converged on the Mall in Washington, D.C., for President Harry Truman's inaugural address. Federal employees had the day off to attend. It was the first televised inaugural, so those who watched in person were joined by hundreds of millions more across the United States and beyond. Truman's twenty-minute address aired on every television network and was beamed into school classrooms. The video was translated and shared worldwide. The world watched, listened to, read about, and discussed his words.

Everyone in Truman's audience had been touched by World War II. My grandmother Mimi considers herself among the fortunate ones. She was one of the few women to attend college in that era, at the University of New Hampshire. After the

Japanese bombed Pearl Harbor, she was called in by the dean. He told her that math teachers were desperately needed to fill teaching roles vacated by servicemen. Mimi majored in math, and though she had a semester of courses left, her professors had determined that she was ready to teach. After writing a paper to satisfy a stickler sociology professor, Mimi earned her college diploma a semester early. In 1942, instead of partaking in the spring semester of her senior year, Mimi was shipped off to rural western Massachusetts, where she taught math and lived by herself.

That same year, her fiancé, John, left to serve his country and never came back. When it was all said and done, he would be joined by 3 percent of the world's population.

By Truman's inaugural, then, people were fed up with world wars, and he spent the first part of his address echoing the prevailing wisdom that Communism posed the greatest threat of another one. Against this backdrop, his first three points were expected. The United States would do all that it could to support the newly formed United Nations, war recovery, and a military alliance between North America and Europe.

Truman's fourth and final point was the surprise. As he put it: "Our aim should be to help the free peoples of the world, through their own efforts, . . . to lighten their burdens."

This was an unprecedented goal for the United States, or for any other nation. Truman was declaring that one way to prevent future conflict was to care for people beyond national boundaries. That must have sounded good to Mimi. She certainly cared for all people. Plus, she was just seven years removed from losing her fiancé. Her new husband worked in defense, and she would have draft-eligible sons before she knew it.

When Truman set this new goal on the world's biggest stage, there was no consensus on how to achieve it. How would we "lighten burdens of the free peoples of the world"? He gave a bit more detail on how he hoped it would go: the United States would help free people help themselves, with "more food, more clothing, more materials for housing, and more mechanical power."

There's a fancy name for this kind of statement. It's called an *anaphora,* one of the oldest rhetorical tools. An anaphora is made by repeating the same word or phrase at the beginning of successive clauses, for emphasis. Whitman used anaphora to emphasize multitudes in "Song of Myself":

> *Have you reckon'd a thousand acres much? have you reckon'd the earth much?*
> *Have you practis'd so long to learn to read?*
> *Have you felt so proud to get at the meaning of poems?*

By repeating "have you," Whitman drew attention to the self. By repeating "more," Truman emphasized adding.

Truman's anaphora of *more* signaled the United States' postwar goals. As he went on: "Greater production is the key to prosperity and peace."

At the time, there was no greater compliment that could be

bestowed upon a social goal. Greater production was how my grandmother's husband and sons would avoid the fate of her fiancé.

Looking back now, it's easy to overlook the significance of Truman's call for more for all. Remember, though, that throughout history, most upstanding citizens—whether at Potbelly Hill, Rome, or Coba—were not in the business of adding wealth. Entrepreneurial traders and merchants could even be social outcasts, or worse. (Just look up *usurers* in the Bible.) Sure, there were oligarchs, feudal lords, and icons of excess like Nero or Marie Antoinette. But most people weren't pursuing economic more.

Truman's televised decree marked a tipping point for an idea that had been gaining steam since the 1700s, when the economic philosopher Adam Smith argued that growth was the fairest way to improve situations for as many people as possible. Without economic growth, most people had no chance to satisfy what Smith considered a natural desire to better our lives: not the laborers who made sun-dried bricks in Mesopotamia, not the Jewish prisoners forced to build Rome's Colosseum, and not the slaves who quarried rock for the Washington Monument.

By 1949, reality had reinforced Smith's ideas about more. Even with the setbacks of world wars, the economic growth brought by the Industrial Revolution had advanced living standards for many. Making a similar point in the opposite way, high-profile programs meant to help people by stifling growth had been spectacular failures. During the Great Depression that preceded World War II, farmers were paid not to harvest their cotton and not to take their slaughtered pigs to market. The hope was that reducing supply would drive up prices.

No such luck. The Depression worsened. At its depth, one in four workers in the United States was unemployed, national income was cut in half, and there were widespread food shortages, no thanks to those wasted pigs. Mimi, who again considered herself fortunate, shared her shoes and her dinner with her twin sister. Things weren't any better elsewhere in the world, which created an opening for Nazis and other fascists.

The failed anti-more programs during the global Depression were fodder for the economist John Maynard Keynes, who, building on Smith's ideas, was overturning the long-standing notion that adding wealth was immoral. Keynes argued that individual consumption was the path to collective prosperity. If more people buy shoes, then companies grow to meet this demand, which creates better-paying jobs for more people, which means more people have more money to spend on shoes, and so on.

In the Depression, Keynes explained, this virtuous adding cycle had turned vicious. People responded to having less money by spending less, which, while logical on an individual level, also shrank the overall economy, which meant that people had even less money to spend. When this global economic crisis led into World War II, the nations involved were eventually forced to spend money on military growth, which broke the vicious cycle.

After the war, Keynes's ideas about individual adding for the collective good began to infuse government policies in Europe and in the United States. Truman's inaugural, then, canonized a patriotic duty to consume and took this same thinking global. The United States backed a system in which everyone could—and should—pursue dreams of economic wealth. More was now a moral objective, the key to peace.

In the aftermath of Truman's inaugural, Brazil, India, Egypt, and Mexico were among the nations that embraced rapid economic adding. If a nation wanted to be part of international trade and finance, they had to commit to growing their economies. Growth was the mandate from the International Monetary Fund and the World Bank, new organizations set up to ease burdens through "more food, more clothing, more materials for housing, and more mechanical power."

There were well-founded and now-familiar counterpoints to this new more-ality: worry that too much growth would concentrate wealth among too few, warnings that endless expansion on a finite planet was physically impossible. But at the time, and for the people to whom Truman was speaking, the way to lighten burdens was through more. For a farmer in India, or Algeria, or in western Massachusetts, more production did, on balance, promise peace. My grandmother bought in, and her four children, seven grandchildren, and thirteen great-grandchildren have so far been spared the horrors of another world war.

It's not just my family; the ideas in Truman's anaphora of more have allowed more people to enjoy life on earth. It took all of history for the human population to grow to the roughly 2.5 billion people alive in 1949. It then took about seventy years to triple that number. Even with three times as many people, global per capita income also grew from around $3,000 per year in 1950 to around $14,500 in 2016. Global life expectancies have risen from forty-eight in 1950 to over seventy today. Around 55 percent of people over the age of fifteen could read in 1950; today, more than 85 percent can. Billions still live in dire poverty, but billions more have choices and opportunities that Mimi did not.

Still. We've learned from evolution and from Maya Lin that it's add *and* subtract. Just because more has made things better, doesn't mean that less cannot.

2.

Food, clothing, materials for housing, and mechanical power are not the only things we've chased since the mid-twentieth century. We have also given ourselves more to do. No matter how judicious your employers' vacation policy, it's likely that medieval peasants got more time off than you do. Using up our time on things we'd prefer not to be doing is bad enough. Even worse, we have come to be proud of this silly behavior. In his classic *New York Times* essay, "The 'Busy' Trap," Tim Kreider helps readers prove this to themselves: "If you live in America in the 21st century you've probably had to listen to a lot of people tell you how busy they are . . . It is, pretty obviously, a boast disguised as a complaint." Seeking other kinds of prosperity, we burn through our most valuable resource: time. Even though we'll never have more of it than we do now.

Whether working or on vacation, we neglect subtraction as a way to improve our days. Recall my team's study of the trip with twelve different activities in twelve different locations in a single marathon day in Washington, D.C. People overwhelmingly added even more things to do. Before my team did that study, but about a year into our work together, Ben sent me an email proudly claiming to be taking our research to heart. He, Gabe, and two of their colleagues had decided to install what they were calling a "no-bell" in the common area between their offices. It's one of those triangle-shaped dinner bells one sees in old Westerns.

The problem with Ben is that he's smart, emotionally intelligent, and physically welcoming, even when he's sporting a thick winter beard. That combination means Ben often has to say no to people who want him to do something for them. Not doing these marginally useful tasks gives Ben time for more meaningful work, like mentoring students and thinking about Legos. To override his instinct to display competence by saying yes to everything, he gave himself the reward of ringing his no-bell.

Ben was rightfully proud of his gimmick, and if he wasn't

always so helpfully skeptical toward my ideas, I may have thought twice about pointing out that saying no is not subtracting. When someone rings the no-bell, it's because they haven't added some new activity, which is not the same as taking away an activity from what they were already doing.

Ben includes suggestions when he points out flaws in my reasoning. In this case, I was able to return the favor. I told Ben that if he really wanted a subtractive approach to productivity, he needed a stop-doing list. These lists had come to my attention by way of the management expert Jim Collins's book *Good to Great*.

With the no-bell, Ben could either add tasks to his already packed workload, or he could add fewer to no tasks and ring the bell for whichever ones he passed on. Ben's current workload is encoded in his mind as an unbreachable baseline, like the square footage of my pre-renovation home, or like a number line that ends at zero. The bell doesn't provide for taking away beyond that point. I hoped a stop-doing list would shift that baseline, cuing Ben to consider both what he might add to his daily regimen and also what he might remove from it.

This is easier recommended than done. In the first half of the 1990s, the sociologist Leslie Perlow discovered some of the first evidence of what we now know is a widespread failure to subtract to-dos. Perlow showed how this failure leads to "time famine," which we can experience at work or on a day trip—whenever we feel like we have too much to do and not enough time to do it.

Perlow first focused on software engineers. She chose this group strategically, skeptical of the media narrative that glorified their long hours. She observed the engineers in their natural habitats: "in their cubicles, in labs, in meetings, and in hallway conversations." She ate lunch with them, attended company parties and happy hours at the local bar. She followed her subjects over nine months that began when funding was committed for their project and concluded with the official launch of the software, even traveling with the engineers on a two-day bus trip to the launch event.

Perlow combined her methodical observations with extensive interviews. She talked with her subjects, of course, and also their coworkers, managers, and members of their families.

She analyzed this data by breaking down her subjects' lives into blocks of time spent on various activities. There were individual work activities, interactive work activities, social work activities, and personal affairs, a category for activities that had nothing to do with work, which allowed Perlow to document engineers making their draft picks for fantasy football.

By carefully documenting how the engineers used their time while they were at work, Perlow found that they did, indeed, have more to do than time to do it in. She also found that many of their obligations—especially the interactive and social activities—were self-imposed. The software engineers attended time-sink meetings and long group lunches not because they were required to but because they felt it would be socially unacceptable not to.

Not surprisingly, the time famine was damaging the engineers' relationships outside of work. Perlow also found that individual engineers feeling overextended was harmful to group output at work. Time famine is not good for morale, it is not good for relationships, and it is not good for business.

Perlow devised a change. She worked with the engineers to designate times when it would be socially unacceptable to participate in interactive or social activities. During these quiet times, there would be no all-hands meetings and no badgering the engineer in the next cubicle to take a working lunch. Perlow's quiet time was a stop-doing, and it wasn't easy. To get buy-in, she had to frame it as "adding" blocks of quiet time. And Perlow's change didn't last long. The engineers quickly reverted to their self-imposed time famine.

The software engineers (and Ben) are not the only ones in the busy trap. A U.S. Army War College report finds that army officers have been caught. The time famine is so strong that it forces these upstanding leaders to be dishonest.

In the most galling example from the report, officers have 256 available days in which to fit 297 days of mandatory activities. Yes, you read that right. It is literally impossible for the officers to do all that is required of them. Their decision is not whether to cut corners, it's which corners to cut.

As the Army War College report observes, "The Army resembles a compulsive hoarder." Just as hoarding used sticky notes and old newspapers is a sign of anxiety and depression, the hoarding of to-dos is psychologically damaging to the military officers. It forces them to act in a way that directly contradicts their hard-earned identity. Following orders is a relatively small facet of Ben's professorial identity, and he still feels guilty when he misses one meeting to attend another. Imagine the mental anguish forced corner-cutting causes for the officers, who have achieved their ranks by doing every single thing that is asked of them.

The officers face this impossible situation because new to-dos have been added faster than old ones have been subtracted. There is simply not enough time for them to do their jobs.

And yet, the recommendation at the end of that Army War College report is to "exercise restraint in the propagation of requirements." But "exercising restraint" would be ringing Ben's no-bell a few times. When the tasks exceed the time available to do them, restraint is not enough. To improve the officers' situation, stop-doing would be required.

Given the situation in the private sector and the military, you won't be surprised that government agencies also fail to stop-doing. The *Code of Federal Regulations* is a record of all the rules made by federal agencies in the United States. The *Code* has ballooned from around 10,000 pages at the time of Truman's speech to more than 180,000 pages in 2020—a rate of growth to make Keynesians proud.

Many of these rules are good to have. It's nice to take Ezra to a diner and not worry about him inhaling secondhand smoke; even better to know that the Clean Air Act protects him, and everyone else, from all sorts of hazardous air pollutants that could be freely spewed in the time of Truman. But a failure to prune outdated rules leaves already stressed regulatory bodies with less time to do their essential work. Neglecting subtraction also leaves them open to critique from those who might profit from fewer regulations.

In his own address to the nation, in 2012, President Barack Obama described how his administration was taking on the issue of too many to-dos. A quote from his State of the Union that struck a chord on Twitter highlighted a subtraction: as he put it, "We got rid of one rule from 40 years ago that could have forced some dairy farmers to spend $10,000 a year proving that they could contain a spill—because milk was somehow classified as an oil. With a rule like that, I guess it was worth crying over spilled milk."

The tweeting was mostly about the unfortunate punch line, which Obama appeared to realize was doomed even as it came out of his mouth. Let's forgive the "spilled milk" and focus on the stop-doing that preceded it. A year earlier, in January 2011, Obama had issued an executive order (#13563). Stipulation 6b in the order was that each federal agency should "review its existing significant regulations to determine whether any such regulations should be modified, streamlined, expanded, or repealed so as to make the agency's regulatory program more effective or less burdensome in achieving the regulatory objectives."

Stipulation 6b looks like one of the cues from my team's experiments. Just as we reduced subtraction neglect in Andy's grids, stipulation 6b cued the agencies to subtract—to consider "streamlining" and even "repealing" regulations. Regulatory programs could change for the better by becoming "more effective or less burdensome."

We know that the cue phrase could have been "more effective *and* less burdensome," to avoid creating a false sense of contradiction between adding and subtracting. But we're not the ones writing executive orders, and, overall, stipulation 6b did a rare and powerful thing. It brought about stop-doings.

Having been cued to think about subtracting by President Obama's executive order, the Environmental Protection Agency (EPA) noted that the regulation classifying milk as an oil was more burdensome than effective. The original regulation, in place since the 1970s, has prevented harmful industrial pollution from spoiling waterways within the United States. Dairy farmers, who depend on these waterways, were happy to do their part containing things like used fuels and leftover pesticides. But the farmers had long questioned whether milk really needed to be subject to the same containment rules as fuels and pesticides. When the executive order prompted the EPA's review, milk was removed from the list of pollutants classified as oils.

As EPA director Lisa Jackson put it: "This step will relieve a potential burden from our nation's dairy farms, potentially saving them money, and ensuring that EPA can focus on the pressing business of environmental and health protection." The step also relieved a burden on Jackson herself, who, despite working with farmers to rewrite the legislation, had been repeatedly forced, including in front of Congress, to refute the notion that her agency wanted to keep this regulation. It *was* worth crying over spilled milk. Savings are projected to exceed $1 billion, not to mention endless hours of agency work that could now be devoted to truly harmful polluting.

Despite their power, stop-doings remain a revelation to many. So here's one final story—my personal favorite—to help you appreciate the broad potential of subtracting in action.

My cousins Chip and Josh and I were all born within three months of each other. Growing up, we convinced our parents to share the same vacation week at a beach house in Ocean City, New Jersey. The last shared summer was the one when the go-kart track opened next to the boardwalk.

The three of us had just turned too old to be riding go-karts, so we spent our evenings doing the same things any adolescent with a soul does when riding go-karts. We bumped each other into the metal guardrails on the side of the track. We peeled out from the starting line before the green light. And we ignored the checkered flag and waving attendants to take one last victory lap with everyone else parked and watching on.

One evening, after the three of us had already been banned from the go-karts for the night, Chip came up with an idea. He had a special gift for mischief: he was the one who figured out that we could acquire a few hundred calories' worth of penny candy using only the "give-a-penny, take-a-penny" stash on the counter by the cash register. That night, he changed his shirt and hat and returned to the go-karts, convincingly disguised, for one last solo ride. Josh and I watched from the boardwalk as he took two innocent-looking laps around the track.

On the third lap, when he reached the farthest point from the starting line, Chip took action. He unbuckled his seat belt, leaped from his still-moving kart, quickly scaled the eight-foot-high fence surrounding the track, and sprinted across the adjacent parking lot, disappearing into the night.

As his empty vehicle rolled to a stop in the middle of the track, Josh and I hyperventilated. The parents expecting to watch their kids circle the track had no idea what to make of the transformed situation. They looked like they had just seen a streaker.

Whether it is Chris Rock setting up a mic drop or Chris Farley suddenly ripping off his shirt, comedy often comes from the unexpected. What my cousin did was completely unexpected, and whether or not I've done it justice here, it's one of the funniest things I've ever seen.

What does this have to do with stop-doings? Well, if the essence of Chip's tour de force was to apply a visible stop-doing to the go-karts, then it was through an unexpected subtraction that Chip brought laughter to Josh, me, and—as the story is passed along in family lore—across four generations of our extended family.

If you don't buy that Chip's stunt was a stop-doing, that's fair. At least you know another way to have fun on the go-karts.

3.

Like Hazel Rose Markus, the University of British Columbia professor Liz Dunn has professor and surfer multitudes. Once, when surfing, Dunn was attacked by a shark. After returning from the emergency room, she described herself to a reporter as "really lucky," because the shark had not bitten her until *after* one of her best rides ever. Her glass-half-full attitude makes sense. When she's not surfing—and probably when she is—Dunn thinks about happiness. And Dunn has found that, when we overlook stop-doings, we not only fail to streamline our schedules, we miss a chance to make ourselves happy.

Thanks to such findings, I've gotten better at subtracting to-dos. I had lots of room for improvement, having been raised by parents who will spend thirty minutes going to pick up pizza, just to save the five dollars to have it delivered.

On top of my upbringing, I face the same mental barriers as everyone else. When we add things, we get something tangible to show for it. When we get rid of a to-do, we get an empty spot on our calendars. As we have just seen, it's challenging enough to stop-doing when it's free. Spending money to save time is a stop-doing that you have to pay for.

Time, it turns out, is worth the investment. In one of her studies, which was led by her student Ashley Whillans, Dunn decided to see what her team could learn from the shockingly small percentage of people who do spend money to save time. They asked more than six thousand people, from North

America and Europe, whether they spent money on time-saving services like cleaning, cooking, and household maintenance. The rare few who did invest in stop-doings reported greater life satisfaction.

Now, my first thought on hearing this was that it made total sense. Of course people who use time-saving services are going to be more satisfied. These are people who have enough money to pay someone to cook and deliver dinner for them.

But, as the researchers found, it was not about the money. Millionaires who paid to avoid the busy trap tended to be happier than millionaires who did not. And it was the same for people living on minimum wage—happiness found those who used some of their scarce income to improve their schedules. Dunn's team had uncovered a convincing correlation between greater life satisfaction and fewer to-do's.

To be more certain that spending money to save time was *causing* the increase in satisfaction, the researchers ran a field experiment. In it, they gave working adults two payments of forty dollars on two consecutive weekends. On one weekend, participants were randomly instructed to spend the forty dollars in a way that would remove something they found unpleasant from their schedule. On the other weekend, participants were assigned to spend the forty dollars to acquire something tangible. At the end of the days in which participants had spent their money, researchers would call participants and ask a series of questions about how they spent their money and time and about how they felt. The researchers found that, after buying time, the participants felt more positive, less negative, and less stressed.

Investing in stop-doings helps us avoid and relieve the time famine that plagued Leslie Perlow's software engineers and those Army officers with more requirements than days. When the threat of personal time famine and busy traps is not enough, I remember the possibilities—what I could be doing otherwise. I'm now more likely to pay a handyman to come hang pictures and fix the porch railing. This is partly because I'm terrible at home projects, but mostly because those are priceless family hours, especially so long as Ezra wants me to go on bike rides with him.

As Band-Aids are to adhesive strips, Strider bikes are to "balance bikes." These are the pedal-less mini bikes that have given preschoolers like Ezra high-speed independence that my generation didn't have until we took off our training wheels. These bikes are propelled not by chains and pedals but by toddlers "striding" with their legs, propelling the bike forward like a Flintstones car.

As a two-year-old on vacation, Ezra hadn't spent more than an hour total on his Strider bike before he was coasting down the Ocean City boardwalk. Using his Strider for his mile-long school commute, Ezra could, on a slight downhill, reach speeds faster than a sprinting father. And Ezra was not special in his riding ability. His cousin and a friend from preschool were even more adept, part of the growing circuit of Strider bike racers who take their tricked-out mini bikes on jump-filled courses.

Strider bikes add a couple of extra years at the beginning of kids' riding careers. What's more, once Ezra decided it was time for his "big-kid" bike, we didn't have to bother with training wheels. He could already balance and just needed to learn to pedal—and then, to brake.

Children's bikes have been marketed as their own distinct class of bicycle for almost a century. There were plenty of design changes over that time: training wheels, fatter tires, shock-absorbing forks and seats, more and more speeds, and contraptions that connect a kid's bike to a grown-up's like a caboose.

It's remarkable, but perhaps not surprising given what we have learned, how long it took for someone to think of subtracting the pedals, which made two-wheeled bikes ridable for a whole new age group—and salable to their parents.

Just as stop-doings face long odds against time famine, subtraction is an unlikely mother of innovation. But Ezra has the unlikely subtraction of the pedals to thank for his extra years of fun.

For Strider bike inventor Ryan McFarland, and for Anna Keichline and her hollow blocks before him, subtracting led to a product that is clearly better. And subtracting is so rare in modern innovation that less need not even be functional. Its novelty alone can be marketable.

Around the time Chip stopped-doing the go-karts, Nike became a billion-dollar company, on its way to dominating the sneaker market and growing into the largest sportswear company in the world. If Chip, Josh, and I weren't already wearing Nike Air sneakers—any of the varieties with the visible air pocket—we were hard at work convincing our parents to buy them for us.

Over a decade earlier, in 1977, Marion Rudy had pitched his air concept to dozens of shoe companies, none of which wanted to follow through with it. As usual, less didn't strike people as intuitively better. Rudy finally worked his way down to Nike, then a small and specialized company. As the story goes, Nike cofounder Phil Knight took a pair of the prototype shoes out for a run, and he liked how they felt. Nike began selling the shoes to their core market of elite runners. Eventually, Nike realized that Air appealed to people like my cousins and me.

For about a decade, Nike's less was invisible. You had to trust that the air was in there. Things really took off when Nike showed what Rudy had done. The iconic Air Max 1s featured a window on the side of the sole—to show the air. The window made the less noticeable, and the noticeable less proved marketable.

Here's Tinker Hatfield, the designer who introduced the Air Max 1, reflecting on this pivotal moment in sneaker history: "People were looking for something different . . . the fact [that the Air Max 1] had the air window in the sole and the frame color around it meant it looked a lot different than other shoes in its day."

Visible less is not the only reason for Nike's success (my cousins and I would have sported Keds had Michael Jordan worn them). And perhaps there is some functionality in that empty pocket. But there's little doubt that the difference Hatfield refers to has helped. Nike's subtraction hit us as a very fresh take on the sneaker.

Ryan McFarland subtracted to extend the joy of bike riding to a new demographic. Tinker Hatfield subtracted to create a difference that surprised and delighted with its freshness. I wondered whether innovators who subtract are, in fact, rare.

Google keeps a database of the patents issued in the United States. These patents recognize, and protect from copycats, one-of-a-kind transformations to products, processes, or machines. I wanted to see whether, with computerized text analysis of the patent descriptions, we might generalize about the changes that led to them. A patent description that repeatedly used words like *add* and *more* would indicate one approach to change, while a description laden with *subtract* and *less* would suggest another.

My speculation was converted into useful insight thanks to Katelyn Stenger, a Ph.D. student at the time, and Clara Na, then in her second year as an undergraduate. Katelyn and Clara first found the eight closest synonyms to *add* and *subtract*. Next, with a bit of help from a text analysis program, they scanned billions of words of patent descriptions and noted every time one of our adding or subtracting synonyms were used. We knew this computerized approach would miss some patents a human might judge to have subtracted, but which did not use any of our subtracting synonyms. That was acceptable because it would be the same with adding.

Some of the patents we uncovered are too good to believe. An adding favorite is:

Multilevel / multi-threshold / multi-persistency GPS / GNSS atomic clock monitoring

A subtracting highlight is:

Backless and possibly strapless brassiere with a reinforcing plate

Yes, you will notice, the reinforcing plate in this "backless and possibly strapless" bra reflects an innovator who was thinking add *and* subtract.

Our results revealed more than just fun examples. Across the more than forty years of digitally available patents, the adding synonyms are used about three times as often than the subtractive ones. Furthermore, the imbalance is widening over time. The use of subtracting terms has remained relatively flat since 1976, while the use of additive ones has nearly doubled. And the patent language did not simply reflect normal patterns of word use. Whereas our subtracting synonyms were far less frequent than the adding ones in the patent text, they are actually more frequent than the adding ones in newspaper text.

Based on this research alone, one might wonder whether adding just tends to be a better way to get a patent. Maybe the innovators themselves were not neglecting subtraction but instead were responding to a system in which adding lots of "multis" to your project title makes it more likely to get approved. That said, the pattern in the patents does look suspiciously like the adding that we observed in itineraries, Legos, and grids—holding people back from better changes.

4.

With these modern adding trends, maybe the basic issue is that less doesn't pay off in capitalist markets. After all, one distinguishing feature of capitalism, capital accumulation, is by definition additive. Perhaps any effort to stop neglecting subtraction is futile because the system is broken. Producers don't thrive by taking away. Dockworker-activist Leo Robinson was a Communist. I couldn't profit by subtracting from the square footage of my house.

By now, we've seen how subtracting can add up, though—in pedal-less bikes, hollow blocks, and even just in the pleasant surprise of being different, like Nike Air. Sue Bierman, those who divested from apartheid, and Liz Dunn's study participants who paid for stop-doings; they all found a way to use competitive markets to continue pursuing the same flourishing desired by my grandmother's generation.

As we shift our focus from understanding why we neglect subtraction to how we can be better at finding less, let's see how our most recent exemplars did it.

Ryan McFarland, for his part, has multitudes right in his Instagram handle: @striderdad. McFarland credits this dual identity for the genesis of his bike epiphany: "The daddy in me wanted to help [my son] succeed; the racer in me wanted to build him a better bike." Hazel Rose Markus would be proud.

McFarland also exemplifies another way to see subtraction: he focuses in on the humans. I don't mean to suggest that the average inventor overlooks humans entirely. It's just that technology, the things the innovators create, ostensibly for humans, tends to get the spotlight.

Here's McFarland again describing how he subtracted to find hidden less:

> When I got to the drive train (pedals, cranks, bearing, chain, sprockets), I realized this was the majority of the weight and complexity. I paused for quite some time at this point as I pondered how to lighten the drive train.

McFarland goes on to describe how his thoughts about lightening the drive train progressed:

> Could I drill holes in it?
> Could I cut away parts of it?
> Until finally . . . could I simply remove it?

Up until here, McFarland is only thinking about the bike.

But, to learn that the answer to that last subtractive question was yes, McFarland considered the humans.

McFarland realized that he was not just trying to transform a bike. He was dealing with a bike, a two-year-old, and what they could do together. Once he considered the human in this way, it was obvious to McFarland, a father himself, that a two-year-old would have plenty of energy to propel the bike forward. And by focusing on the human, McFarland found that the two-year-old could also provide the balance.

It typically takes at least four years on earth to develop the strength and coordination to pedal, but with pedaling removed, balancing alone turns out to be quite doable after two. Balancing on two wheels is an unexpected skill for the two-year-old demographic.

After Ezra whizzed around his great-aunt's convent, she kept insisting I explain to her what technology in his Strider bike allowed it to automatically stay upright. Engineers have spent ages trying to build machines that can walk and balance like a human. Even octogenarian nuns, it seems, have gotten so used to focusing on the object, giving credit to the technology, that Ezra's great-aunt thought that the balancing force must be in the Strider bike. But the invisible force was in her grandnephew. Judging by the serious and accomplished look on Ezra's bike-riding face, he was proud of it.

As with the notion of more as moral, the promise of applied science was gaining steam at the moment of Truman's inaugural address. Growth was the way to a better future, and so was technology. The Manhattan Project to develop the atomic bomb had helped bring an end to World War II.

Truman encouraged his worldwide audience to direct new

technology toward development and rebuilding. "We should make available to peace-loving peoples the benefits of our store of technical knowledge," he said, inviting other countries to "pool their technological resources in this undertaking." He boasted that "our imponderable resources in technical knowledge are constantly growing and are inexhaustible." He even linked the vision of more to his love of technology: "The key to greater production is a wider and more vigorous application of modern scientific and technical knowledge."

I don't disagree with any of what Truman said, so long as we define technology as doing things with science. In that case, the atomic bomb is technology, and so is Leslie Perlow's quiet time. However, when we misinterpret technology as just the machines we make, we can lose focus on the humans.

When we refocus on what matters, removing the pedals of a starter bike becomes obvious. One day, running alongside Ezra as he cruised down our street on his, I told him, "You're lucky. We didn't even have Strider bikes when I was growing up."

His unironic response was, "Didn't you have wrenches?"

We had plenty of wrenches. But before Ryan McFarland came along, no one had taken one to a bike pedal. McFarland did it first and, at around one hundred dollars of the best money you will ever spend each, he has since sold more than two million Strider bikes and counting. His nonprofit has donated millions of dollars' worth of cash, products, and time to get even more kids on bikes. Sure, capitalism can reward adding. But, as McFarland reminds us, it's add *and* subtract; there is plenty of profit in subtraction. Especially when we consider the humans behind the money.

From monuments and home renovations to patents and our precious time, it's the same basic story. Economic forces reinforce biological and cultural ones, and eventually, only one in sixty of the participants in our studies subtracts to improve the Lego structure.

Reinforcement to add may be powerful, but it is not irreversible, and it can work for less too. As we have seen, certain cues and views can make us more likely to think of subtraction. The more we think to subtract, the more likely we are to get to less, the more those pathways in our brains are stimulated, the more we will think to subtract, and so on.

In the meantime, our instincts to add, cultures of more, and economies of growth make it so that we have much to gain from taking away. Lisa Jackson's milk subtraction benefited from a legislative process that neglected removal. The stop-doings Leslie Perlow recommended and Chip pulled off were more effective because of their novelty. And whether it's a pocket park in Harlem, a square in Savannah, or Maya Lin's Vietnam Veterans Memorial, the impact of a physical subtraction is enhanced by the adding field that surrounds it. When everyone's adding, it pays to hone the opposite skill.

We now recognize and respect the challenge of subtracting. Next, let's do it.

PART II

Sharing Less

5

Noticeable Less

Finding and Sharing Subtraction

1.

"I didn't have time to write you a short letter, so I wrote you a long one." Mark Twain often gets credited with this quip, and while there isn't a record of him actually saying it, there is no shortage of similar sentiment. Those who admitted to writing more because they knew less would take longer include the statesman Cicero, the scientist Pascal, the essayist Thoreau, and the statesman, scientist, and essayist Ben Franklin. My favorite version—because it has a physician who also enlightened the world questioning his own effort—is John Locke's: "But to confess the Truth, I am now too lazy, or too busy to make it shorter."

Whether in writing or building, ideas or things, we all face situations in which it is easiest or most practical to leave well enough alone. Herbert Simon, who would earn an economics

Nobel Prize for his effort, found that the tendency to stop at good enough was widespread. Simon named this tendency "satisficing," a portmanteau of *satisfying* and *sufficing*.

As Simon discovered, we satisfice because improvements that are possible in theory can be too hard, not worth the effort, or unnecessary. In those cases, imperfect satisficing makes perfect sense. It is the quickest path to a goal. When grocery shopping, I buy the first jar of pasta sauce I see that does not have meat, costs less than five dollars, and will not obliterate my attempts to keep Ezra's daily sodium intake within the recommended range. Sure, I could spend more time and find a jar that is healthier and costs less, but I have moved on to figuring out which noodles to buy.

Stopping at good enough protects us from wasting effort, but if we are not careful, the same tendency can prevent us from subtracting when the effort would pay off. This isn't just the case in a run-on letter from Mark Twain. Think back to our starter bikes. A standard bike with pedals has fewer parts than one with training wheels, but there is no way for a new rider to balance. This is pre-satisficed less—let's channel John Locke and call it "lazy less." A bike with pedals and training wheels is good enough, because a new rider can stay upright on the bike. But Ryan McFarland's version, the Strider bike, goes beyond good enough, helping the new rider more effectively. We are interested in this less beyond more: a post-satisficed less.

Getting to post-satisficed less requires more steps. Then, even if we put in the effort to go beyond good enough, we still

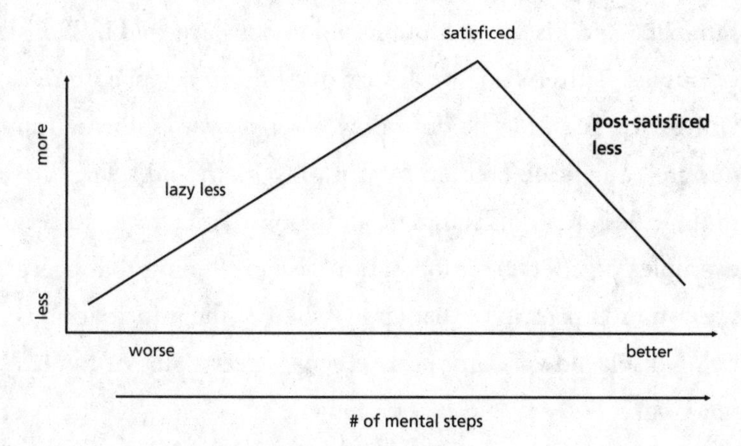

Figure 8: The longer path to post-satisficed less

face all the familiar anti-less forces, from our tendency to overlook subtraction, to our instincts to add, to a society built on the gospel of more-ality. In other words, extra effort can bring post-satisficed less, but so long as we fail to subtract, extra effort will bring post-satisficed more.

We can change that. Having investigated our adding world, we now understand how we neglect subtraction, probably at a level that Franklin or Locke never did. Armed with this awareness, we can move on to finding balance. We are missing chances to subtract; so, how do we find them? And how do we get others to buy in?

Edward Tufte finds and shares less, profitably. A couple of hundred thousand people have paid for Tufte's in-person seminars, and his five self-published books have sold millions of copies. Tufte was a professor of political science at Yale University, but what he does now is reign over "information design," a field he created by bringing math and science to graphic design. In his seminars and books, Tufte deconstructs examples of effective information design, finding that there are common principles that apply whether the information is being displayed via computer screens, street signs, or figures in a book.

With these time-tested principles for sharing information, Tufte helps others go from good enough to the polished less that's beyond it.

Some of Tufte's principles identify the adding problem. He coined the term *chartjunk* to call attention to all the distracting and at best useless marks that tend to find their way onto graphics. Chartjunk calls out third dimensions for two-dimensional data, extra grid lines, overemphasized axes, and less important labels and numbers given equal weight as more important ones. The figure below is a revised version of the first one in this

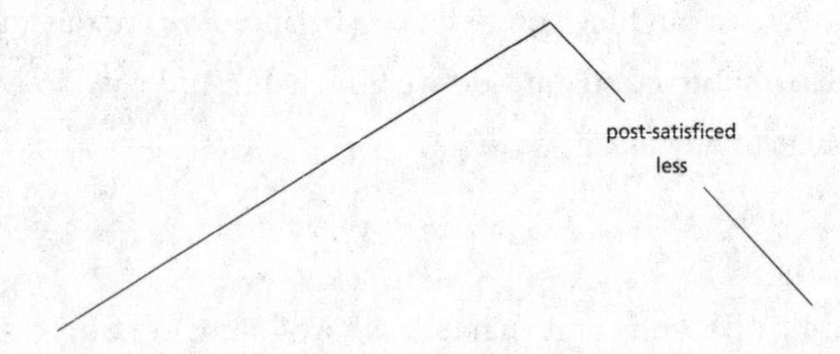

Figure 9: Post-satisficed less: with chartjunk removed

chapter and (hopefully) an example of how removing chartjunk adds clarity.

Subtracting chartjunk is one way to pursue Tufte's larger graphical goal: to maximize our information to ink ratio. Tufte's ratio is a handy way for anyone to assess and revise their graphics, encouraging a more balanced use of adding and subtracting. As we strive to maximize the ratio, we can add information *and* we can subtract ink, or as Tufte puts it: "Erase non-information-ink. Erase redundant information-ink. Revise and edit."

When we maximize the data-ink ratio, the information we share can do more than meet expectations. It can bring unexpected insight. Tufte invokes Maya Lin's Vietnam Veterans Memorial as an example; Lin's insistence on engraving the names of the deceased chronologically (instead of alphabetically) shows the war's scope and evolution—without requiring more etchings on the wall.

It takes more effort to create post-satisficed shorter writing, and it's the same for graphics that maximize information to ink. But the effort is worth it, which is why Tufte's tips are taught to aspiring designers, artists, and scientists; and why *Bloomberg* has called Tufte the "Galileo of Graphics." Since Tufte dabbles in the subtractive art of sculpture, I think of him as *The New York Times* does, as the "Da Vinci of Data."

I think of Kate Orff as the "Tufte of Things." Whereas Tufte teaches us about post-satisficed subtracting in information, Orff shows how to do it in our landscapes.

Just as it takes more mental steps to think of subtraction,

it takes more physical steps to do it. To subtract a redundant phrase, we need to write it first; to clear a vacant brownstone in Harlem, someone needs to build it first.

And to reveal a river, someone needs to cover it first.

Heading into the summer of 1833, Lexington, Kentucky, was home to about six thousand people. By fall, more than five hundred of them had died from cholera. The fortunate died fast. Others hung on for days, their brains aware of their dehydrating bodies. Bodies piled up faster than they could be buried. Orphaned children wandered the streets begging for food.

Like many cities, Lexington grew up around a river. The Town Branch Creek gave the city life, providing a steady supply of water for drinking, growing crops, and running mills. When the Town Branch flooded, however, its water mixed with human excrement from outhouses and animal excrement from free-roaming pigs and cows. Because Lexington rests on porous limestone, the aboveground floodwater cesspool seeped into the underground water that supplied the city's wells. Lexington's decimation by cholera could have been prevented had the Town Branch Creek not flooded the city.

People set about controlling the Town Branch's trip through Lexington. They carved and hardened channels to direct the water and contain surges. Bit by bit, they covered the Town Branch Creek with buildings, factories, and roads. In the few places the water remained exposed, it was indistinguishable from a drainage ditch.

This was the situation in Lexington as the city entered the twenty-first century, with the population approaching three hundred thousand and city leaders looking for ways to improve their downtown. They held a design competition in 2013, offering enough prize money and prestige to attract submissions from large planning firms specializing in city revitalizations.

The winner was a surprise: SCAPE, Kate Orff's tiny firm. Thanks in part to how they are transforming Lexington, Orff and her firm do not fly beneath the radar anymore.

Orff's plan for Lexington is called Town Branch Commons, a public space following more than two miles of the creek's historic path through downtown. Like the post-cholera improvements, Orff's project will control flooding and even filter the Town Branch water. The project will deliver green space and a multiuse trail connecting downtown Lexington to the surrounding horse country. The soothing beauty of water will be returned to downtown, with strategically placed pools and water windows cut through the limestone.

Orff's work is a study in post-satisficed less. On the path to good enough, Lexingtonians had added sewers and channels to control the Town Branch, and then covered it with roads and buildings. This happened everywhere: Minetta Brook remains hidden beneath the streets of Greenwich Village, a few miles south of Collyer Brothers Park in Harlem. Islais Creek flows under Sue Bierman's San Francisco. Covering these waterways improved sanitation and provided valuable real estate. There were unintended consequences, like degraded habitats and downstream flooding—but in stopping cholera from decimating cities, adding satisficed.

To improve twenty-first-century Lexington, however, Orff's plans add *and* subtract. She removed concrete to create the multiuse path. She hewed out limestone to make the pools and water windows. By subtracting, Orff's design not only meets expectations, by preventing downtown flooding; it exceeds and therefore resets expectations, by reconnecting the people with the land and water.

Compared to built-up Lexington, Town Branch Commons may appear natural and even effortless. The new park at the west end of the commons might be mistaken as an oasis set aside at the genesis of Lexington, like Central Park in New York. The sinuous green space drawing people through downtown could seem, like Savannah's squares, to have been carefully integrated from the beginning. But these apparently lazy less spaces are, in fact, subtractions. The less is a result of Kate Orff's vision and effort.

To get to post-satisficed less, Orff and her team had to do more. They had to think about pipes and pumps and concrete and all the other tried-and-true ways to control water. They also had to see the field. They had to think about Lexington's unique limestone geology, its rural bluegrass surroundings, and, like Ryan McFarland, the humans—both in the city's present—and its future. Orff surely sacrificed short-term profit to consider all of this context. For a professional designer, more effort on a competition entry means more unbillable hours. But while good enough works for my pasta sauce, it was unacceptable for Kate Orff's Lexington.

The extra thinking was not in vain. When it came time to fund the construction of Orff's winning design, post-satisficed less paid off. The plan quickly attracted more than $20 million in federal grants, $7 million from the state of Kentucky, and $12 million from local sources. With funding secured, Lexington's physical transformation commenced in early 2020.

Orff herself, and her firm, are now in high demand. Since winning the design contest for Lexington, she has designed a waterfront greenway in Brooklyn, removing roads and restoring natural systems to protect New Yorkers from the next superstorm. Across the country, in the Bay Area of California, Orff is "unlocking" Alameda Creek, removing human-made barriers so that the creek can, once again, carry sediment to nourish protective tidal ecosystems.

To get to post-satisficed less often requires that we have already added, whether by channeling a river, formulating an argument, or cluttering a graphic. It's important to acknowledge this—that we're not starting from zero—because adding first erects a mental obstacle to less. When we see that something has already been done, we tend to leave well enough alone. Whatever is there must be either necessary or too much trouble to reinvent. Or, as countless satisficers have said: if it's not broke, don't fix it. Adding first is enough to turn John Locke lazy; it takes effort to get to less.

2.

Getting there is just half the battle. We also need other people to notice our post-satisficed less. We need our audiences and clients, friends and family, to appreciate that we have made a change that is, in fact, better. Noticeability is how Kate Orff can win a design competition by proposing to *remove* infrastructure from downtown Lexington. Noticeable less is how you create Tufte-esque slides that your audience won't mistake for lazy.

Good writing illustrates this principle. Experts, examples, and research all suggest the same thing: less is objectively better. This is what Twain (maybe) and his predecessors acknowledged. This was the essence of Ernest Hemingway's confessed practice of taking out parts of his short stories on the "theory that you could omit anything . . . and the omitted part would strengthen the story." Less is better is what the research data reveals in Daniel Oppenheimer's "Consequences of Erudite Vernacular Utilized Irrespective of Necessity: Problems with Using Long Words Needlessly." And it is hard to make it through college without reading, or at least being assigned to read "Strunk and White." Compiled over decades of teaching English, William Strunk Jr.'s writing guide was updated in 1959 by his former undergraduate student E. B. White. Their resulting book, *The Elements of Style,* still appears on more course syllabi than any other book.

Perhaps the most famous Strunk and White advice is their blunt reminder to subtract: "Omit needless words."

Given all this conventional wisdom, Gabe, Ben, Andy, and

I thought writing was one context in which we might observe people subtracting. Alas, whether changing their own summaries or someone else's, our participants tended to add. They seem not to have internalized Strunk and White.

Or maybe our participants knew those best practices but figured their evaluators would not notice what had been omitted. Research has shown that an argument's length is often used as a proxy for its quality. Maybe our participants had heard that longer responses on standardized tests are reliably graded higher. Just because more concise writing has been judged objectively better, we were reminded, what really matters is how your audience responds.

Courtney was among the students to enter the design contest that led to our home addition. A couple of years later, she was applying to graduate schools and shared with me an application question she knew I would find amusing. One of the questions asked by Harvard's Graduate School of Design was:

> What is your reaction to the phrase, "Less is More," an aphorism found in many disciplines? (300 words)

What fascinates me in this question is the word quantity looming at the end. Was three hundred words the maximum length Courtney was allowed? The minimum? An optional target? Did whoever had written the question intend the irony of suggesting a quantity on the question about less is more?

It's fun to find fault with Harvard, but teachers everywhere encourage us to add. History professors specify that an essay should be "at least ten pages." Math instructors take off points if you don't "show your work." Even when our teachers don't explicitly ask for more, we provide it. Who among us hasn't left in a part of an essay, design, or calculation not because the part improved what we turned in but because we wanted someone to see the work we had done?

To submit a response much fewer than three hundred words, Courtney would have to overcome that evolutionary instinct to show competence, to prove to others that she can shape her world. What's more, she could safely assume that nearly all her fellow applicants would be writing around three hundred words. After all, three hundred words is visible evidence that you put in the time and thought. The long response will not be mistaken for lazy less. Courtney was determined, though. She staked her competence on a fifteen-word haiku:

> A wall torn down or; a single door that opens; can mean so much more.

Whether in school or practice, there is comfort in providing more. When a planning firm outlines a vision for Lexington with sleek new buildings, wider roads, and tiny data-collecting computers embedded in both, they reassure whoever is paying that bringing in the professionals was a smart move. When Kate Orff subtracts to highlight Lexington's natural beauty, she risks appearing unthorough. She puts in more effort and has less to show for it.

So, what do you do if you have gone beyond satisficed and people aren't responding to your less? How do you make what's not there undeniable? How do you not only subtract words but get fans and critics to applaud it?

The Hemingway of New Jersey, Bruce Springsteen, refers to *Darkness on the Edge of Town* as his "samurai record, all stripped down for fighting." Remember, to "strip down," there needs to be something there in the first place, whether it is a rock and roll song, a built-up city, or a chartjunked graphic.

Springsteen subtracted words to get to his *Darkness* lyrics. On his previous albums, he had proven more than capable of meandering vernacular. Here's the opening line to "Blinded by the Light," the first song on Springsteen's debut album, *Greetings from Asbury Park:* "Madman drummers bummers and Indians in the summer with a teenage diplomat."

Compare that mouthful with the opening line to "Racing in the Street," a track from *Darkness* (and Ezra's lullaby): "I got a '69 Chevy with a 396."

Just as he shortened long words on *Darkness,* Springsteen omitted needless ones. Songs on the album average around 225 words each, far fewer per song than on any of his previous albums.

Darkness also has stripped-down instrumentals. As Springsteen describes in his autobiography, "When the drums are forceful but moderate, they leave room for a big guitar sound. When the guitars are powerful but lean, you can have drums the size of a house."

Not only do the individual songs on *Darkness* achieve a stripped-down sound, the album as a whole also exemplifies persistent subtraction. Springsteen whittled more than fifty recorded tracks down to just ten. Some of those he cut became hits for other artists: Patti Smith got to #13 on the charts with "Because the Night." Gary U.S. Bonds peaked at #11 with "This Little Girl." The Pointer Sisters hit #2 with "Fire." The audacity to carve those hits from his samurai record is even more impressive when we consider that, at the time, he had yet to have a top-20 single of his own.

Springsteen sacrificed words, sounds, and songs. But, because he was so persistent with his subtraction, his less is noticeable. Writing in *Rolling Stone,* the notoriously hard-to-impress rock critic Dave Marsh called *Darkness* "nothing less than a breakthrough," an album that "changes fundamentally the way we hear rock and roll, the way it's recorded, the way it's played." It was named 1978's album of the year by the prestigious *New Music Express* and remains high on any self-respecting list of the best rock and roll albums of all time. The stripped-down aesthetic has inspired the rock spectrum—from grunge's Pearl Jam, to alt rockers Rage Against The Machine, to new wavers the Killers.

One need not be a rocker or critic to appreciate Springsteen's samurai record. Even before I knew why *Darkness* was different, I knew that it was, and a lot of fans agreed. The album backdropped a 115-show tour that finally got Springsteen out of debt, en route to his selling tens of millions of albums and tickets ever since. Even now that he has more than three hundred songs to his name, his concert set lists are likely to include at least half of the ten tracks from *Darkness.* (Probably "Badlands," "The Promised Land," "Prove It All Night," and the title track. Plus, if you're lucky, Ezra's lullaby.)

Like Springsteen, Maya Lin persisted to noticeable less, getting her anonymous design seen in a pile of more than 1,400 entrants. Kate Orff persisted too, refining her intricate network of pools and water windows and bringing visible delight to downtown Lexington.

In 2017, the MacArthur Foundation awarded Orff with one of their aptly named Genius Awards for exceptional creativity, noting that Orff had extended "the boundaries of traditional landscape architecture." Springsteen's persistent subtraction fundamentally changed how we hear rock and roll. Orff's noticeable less extended the boundaries of a discipline. Subtract enough and, eventually, what's not there becomes the story.

One of my attempts at noticeable less is on the cover of this book. I hope the downward sloping line reminds you of the persistent subtraction that can take you from satisficed all the way to noticeable less. Because the design is mostly the work of a professional cover designer, I also hope it reminds you that noticeable less often requires asking for help.

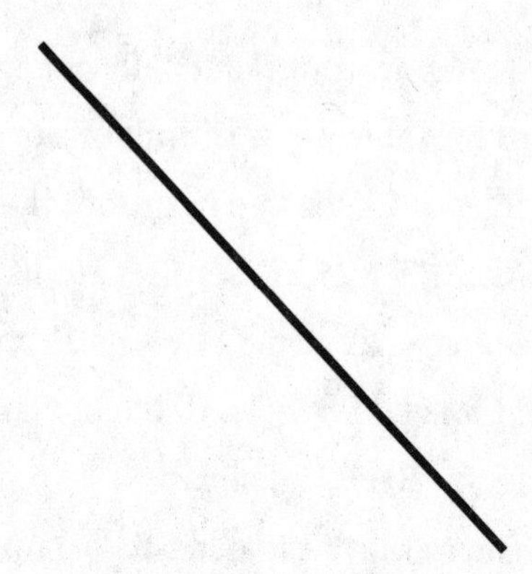

Figure 10: Noticeable less (with the help of an editor)

There's no shame in getting a second opinion. When we create something, whether a window cut through limestone, a *Darkness* outtake, or a paragraph, it's natural to be attached to the work we've already done. Even if our prior work is irrelevant to the decision at hand, getting rid of it makes us feel like we've wasted the effort used to get there. This is why, when my team asked participants to improve their own writing, they were even less likely to subtract words than when the writing was someone else's. Rare are the Courtneys who can send Harvard a haiku.

The rest of us benefit from help, ideally from someone with little attachment to our prior work, and maybe even someone who is professionally better at finding less. Ernest Hemingway worked with Max Perkins, an editor who also harnessed the brilliance of F. Scott Fitzgerald, Thomas Wolfe, and Marjorie

Kinnan Rawlings. The Strunk and White pairing worked because White was an editor.

Most everything we pay to read benefits from editors finding the less we can't get to on our own. Even after I put every word of this book on trial for its life, an editor read tens of thousands more of my words than you are reading here. I can assure you that your experience is better because I got help.

3.

As my team made sense of the adding in our early studies, I convinced a dozen students to sign up for the first version of my course on subtracting. On the first day of class, after introductions, I tried to explain what these pioneering students could expect to learn. I started with an example: Ryan McFarland and the long wait to remove pedals to improve kids' bikes. To show that subtraction gets neglected in many situations, I shared highlights from the studies with Legos, Andy's grids, and patents. I had next planned to outline the overlapping forces behind this behavior, like our instincts to add and show and our more recent more-ality. But before I could get into these forces, not five minutes into my overview of the course, there was a hand in the air. I called on the student, Sarah, whose question was more of a translation for her classmates:

"You mean, kind of like Marie Kondo?"

As I have since learned, the tidying mogul is famous. There

are those, like Sarah, who got on the Kondo bandwagon early. These devotees have clean kitchens and very few T-shirts, which are rolled instead of folded, as Kondo advises in her book *The Life-Changing Magic of Tidying Up*. Others came to Kondo after binge-watching her post-book Netflix series. Monica was part of this second wave, which is why it is easy to find things in our coat closet.

Then there are those who reluctantly admit to having heard of Kondo but claim not to have read her books or watched her show. These people, especially common among professors, tend to resist any form of prescriptive self-help, especially that which is not derived from scientific evidence.

Despite my long-standing obsession with less, I had remained part of this too-good-for-Kondo faction well into my work with Gabe, Ben, and Andy. But questions like Sarah's convinced me to see for myself what she had to say.

As it turns out, not only does time spent with Kondo spark joy, but much of her home-tidying advice is scientifically sound. Her lessons are not grounded in research; there is none of the systematic rigor and control that distinguishes modern science. Kondo does not pretend that there is. Her tone, observations, and advice are spiritual. And yet, through trial and error in one specific context, Kondo has derived tips also suggested by the science. With witticisms like "The best way to find out what we really need is to get rid of what we don't," Kondo cues her disciples to think of keeping *and* getting rid of stuff as complementary ways to improve. An entire section of *Life-Changing Magic* expands on her insistence to get rid of the things "intensely and completely," laying out the step of persisting to noticeable less. For me, the most precious Kondo quote is "Tidying your physical space allows you to tend to your psychological space," connecting ideas and things in a way that would please Emerson.

Those are my interpretations of Kondo's teachings. You may disagree, and that's just fine. Her most valuable lesson for us is in how she gets people to do more and end up with less.

What makes Kondo's message unique and powerful is her emphasis on "sparking joy." Whereas default home organization advice is to get rid of things you don't want, or that don't fit, Kondo flipped that logic around—and focused it on us humans. She said we should keep what sparks joy and get rid of everything else. There are a lot of T-shirts, kitchen appliances, and Lego sets that we've never considered throwing out but that certainly don't bring us true happiness. Kondo insists. If it doesn't spark joy, it's time to let it go from your life.

Kondo's spark joy mantra resonated so much that it became a movement and (inevitably) a meme. And, of course, the title of her next book. To persist in our subtraction, and to get others to buy in, it helps to make it fun.

Fun need not be easy. In fact, the challenge of subtracting brings great rewards.

When Ezra was an infant, he had a mobile of stuffed fish hanging above where we changed his diapers. Ezra cried when being changed, until his parents started batting his fish before laying him down. He would stay distracted from crying, excitedly pumping his feet and arms in sync with the swaying fish. As he grew, he could eventually bat the fish himself. Then he could reach high enough to grab them. Finally, on one unforgettable changing, Ezra ripped his favorite orange fish right off the mobile.

If I had to choose just one thing to remember from Ezra's first year of life, it would be the look on his face after he ripped down the orange fish. This was not the "praise me and take a picture" face. This was his "I am serious and capable" face that also comes out for bike riding, Lego building, and ninja training.

I was beaming as Ezra held his fish friend, but he wasn't looking at me for affirmation. He might as well have been the only person in the world. He had shown to himself, in a whole new way, that he could change it.

When we show others that we can effect change in the world, that is showing competence. When we show *ourselves* that we can change the world, that is called *self-efficacy*. Our self-efficacy is our belief in our ability to shape our own motivation, behavior, and surroundings. When we have high self-efficacy, we believe that we can transform ideas and things. With low self-efficacy, we see situations as beyond our control, in which case, why yank on the orange fish, get back on the Strider bike, or make any other attempt to change things from how they are to how we want them to be?

The more I see Ezra's self-efficacy face, the surer I am that there will come a (bittersweet) time when Ezra no longer relies on Monica and me to improve his situations but instead improves them for himself and for others. With any luck, Ezra will find joy in the process of doing so.

One reward for those who persist in the process of improving is that this process can bring what happiness researchers call "flow" states. You are in a flow state when you are so immersed in what you are doing that time passes imperceptibly; all of a sudden, there is only a minute left in the game, the disc jockey announces the last song, or you have reached the last step in the directions for the Lego castle.

In his authoritative book *Finding Flow,* the psychologist Mihaly Csikszentmihalyi lays out a convincing case that these states represent optimal mental experience. It happens, he argues, at the alignment between a challenge and our ability. Put a high school soccer player in a game with Ezra's Maroon Lions, and they won't be challenged. Put them in the middle of a professional game, and they will be too overwhelmed to find flow. To find flow, we need to stretch ourselves, but not in vain.

Getting back to subtraction: lazy less doesn't spark flow because it is not a challenge. There is no transformation.

Adding to get to good enough is slightly more challenging. But satisficing, by definition, doesn't test the limits of changes or of our ability to make them. Good enough does not bring flow.

Flow happens when we go beyond good enough. Admittedly, post-satisficed adding can also bring flow. But subtracting might have a slight advantage in doing so. To see how, consider the author Stephen King's observation that "to write is human, to edit is divine." Or Maya Lin's Springsteen-like version: "My goal is to strip things down . . . I like editing."

Adding a worthwhile new phrase can be an overwhelming challenge, with trillions of possible word combinations. The higher our standards, the higher the risk that the challenge will exceed our ability, that we will feel like the high school player in the professional game. Writer's block is most certainly not a flow state. On the other hand, when we ponder how we might take away from what we have written, our mental search is bounded by what exists. There are lots of possibilities, but at least they are right there on the screen or paper in front of us. Taking away words to transform what's already there may not be our first instinct. But editing is a challenge that matches our ability. That kind of challenge can be divine.

It is telling that our subtracting exemplars all appear to find joy in the process of transformation. Springsteen is in his sixth decade of writing and performing, and he says the last job he had was as a fifteen-year-old lawn boy. Orff is a tenured professor at Columbia University. She could shutter SCAPE tomorrow and still provide for her family. When Kondo embarked on her Netflix series, she had already made millions from her books. And Tufte's political science career was paying the bills. He spent a decade exploring uncharted intellectual territory, with little to show for it. When publishers were not excited about the book by a political scientist about information design, Tufte took a second mortgage on his home to share his data-ink ratio with those who might use it. For Kondo and Springsteen, Orff and Tufte, external rewards seem secondary to joyful flow.

4.

If sparking joy and finding flow are too touchy-feely for you or your audience, there's a more detached way to sell your subtracting. Invert it.

To see how this works, we need to appreciate that the idea of subtraction has a negative *valence*. In chemistry, valence refers to an elemental force that is not necessarily visible but helps explain the elements' behavior. Back in the 1930s, the psychologist Kurt Lewin borrowed this notion of invisible, and influential, forces to help explain human behavior. Lewin defined psychological valence as the intrinsic attractiveness (positive valence) or averseness (negative valence) of an event, object, or idea. As in chemistry, valence is now an established and useful construct for psychology.

One recent use has been characterizing the valence of individual words, so that computers can interpret text closer to how humans do. Word valences are determined by asking thousands of people to classify thousands of words as positive, negative, or neutral. When these answers are averaged, most words come out as neutral. Fewer than one in five words has a negative valence; even the word *less* has a neutral valence. As you might have guessed, though, the word *subtract* is viewed negatively.

Just by reading this book, you are changing your personal valence for subtracting. But what about your customers, audiences, and friends? With any luck, the ideas in this book will permeate culture to such an extent that there becomes a positive, or at least neutral, valence around taking away. Until that happens, inverting subtraction is a shortcut you can use as you try to get others to share your newfound appreciation for less.

Kate Orff uses four verbs to describe her winning design for Lexington, Kentucky. These verbs earn the largest font on Orff's drawings (like the one below) and are used to organize the written specifications for the design. None of these verbs is *subtract*, but three of them could be.

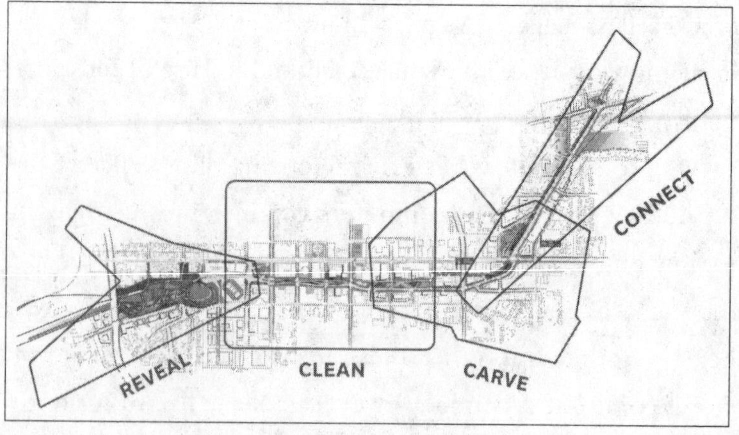

Figure 11: Kate Orff's positively valenced verbs.

Reveal is how Orff describes the westernmost part of Lexington's transformation, the Town Branch Park. Here, the creek is daylit and new open space "gestures" to the nearby bluegrass region.

Clean is applied to the change just to the east of the Town Branch Park. Here, in the heart of downtown, the emphasis is not on lost pavement and subtracted car lanes but on a series of water filtration gardens that "shape an expanded streetscape."

East of downtown, *carve* is Orff's chosen unit of transformation. Parking lots are subtracted and turned into a plaza with scattered pools and windows cut through the limestone to show the water below. (Orff's fourth verb is *connect,* to describe a link between the downtown commons and another trail system.)

It is a lot of semantics. But as Kate Orff seems to know, words matter. Their valence affects how your audience feels. And as it turns out, when it comes to subtracting (or revealing), these valence stakes are especially high.

In 1979, a few months after Herbert Simon received his Nobel Prize for showing that people stop at good enough, Amos Tversky and Daniel Kahneman published a paper demonstrating that we value things we have even more than things we do not. Losing one hundred dollars, Tversky and Kahneman showed, feels more disappointing than gaining one hundred dollars feels satisfying. They called this finding—that the response to losses is stronger than the response to gains—"loss aversion." Loss aversion has gotten a lot of deserved attention, often as exhibit A for behavioral economics. And the

Simon-esque edit to economic thought has also been amplified by Kahneman's own Nobel Prize in 2002, by his bestseller *Thinking, Fast and Slow,* and by Michael Lewis's subtractively titled account of Kahneman and Tversky's transformational ideas: *The Undoing Project.*

To study how we value gains and losses differently, Kahneman used brilliantly simple experiments. Like many others, I re-create bits of his studies in my classes.

The simplest version is to randomly give half of the students a physical object that has some value, like a university-branded coffee mug if you want to be true to the original experiment, or a pencil once you have tired of schlepping coffee mugs from the bookstore to class.

After half of the students have received a pencil, I ask them to write down the lowest price for which they would sell it. Those without pencils I ask to write down the highest price they would pay to get one. The sellers tend to price the pencil about twice as high as the buyers. Since the only difference between the two groups of students is whether or not they recently got a pencil, that must explain the difference in how much they think pencils are worth.

Similar loss-averse behavior occurs with other goods and in different populations (including capuchin monkeys). Brain imaging confirms that losses and gains stimulate different circuits in our brains.

Those who sell things use our loss aversion to their advantage. Car dealers urge us to take that no-strings-attached test-drive, because the more we feel like we have the car, the more value we assign to it. Amazon.com gave me unlimited free two-day shipping for a year. I wasn't going to pay the annual fee to get the service, but I now pay that same fee not to lose it.

You might see loss aversion in situations we've discussed: when San Franciscans had the Embarcadero Freeway, they assigned it more value than an open waterfront. Now that San Franciscans have an open waterfront, they want to keep that.

Loss aversion is powerful, widespread, and well publicized. But loss aversion should not excuse our subtraction neglect. The subtraction we are after is an improvement—and improvement is not a loss, even when it comes via less.

Less is not a loss. But to avoid any misunderstanding, we can invert subtraction—just like Kate Orff.

Her verbs *reveal, clean,* and *carve* are gentler alternatives to *subtract.* They do not invoke a negative valence, and they do not activate loss aversion. If Lexington residents view Orff's subtracted car lane as a loss, then the gained multiuse path doesn't just need to be objectively better than the car lane, the path needs to be so much better that it also overrides the disproportionate sense of loss. But if Lexington residents view the removal not as a subtraction but as a *clean,* taking away is competing on a more level playing field.

Once I began to look for it, inversion was everywhere.

Tufte doesn't tell us to subtract what we've already drawn. He challenges us to "maximize our information to ink ratio."

Kondo inverts subtraction too. Just as giving my students a

pencil makes them value pencils more, when Kondo advises, "Before you start, visualize your destination," she's giving me a decluttered living space, at least in my head. With that tidy vision in mind, a Lego room that sparks joy, rather than the toys I'll need to part with to achieve it, becomes the loss that looms larger than gains.

In Orff's work, inverting subtraction is so pervasive that one wonders whether it is intentional marketing or a way of thinking about change that has become intuitive to her. My guess was that it was subconscious, while Natalie, a student in my subtracting class, thought it had to be intentional. To see who was right, Natalie spent the semester reading and rereading Orff's book *Toward an Urban Ecology.* Natalie scanned all the words used to describe designs, coded those that were additive or subtractive, and counted the number of times each was used.

In her book, Kate Orff overwhelmingly uses words that avoid loss aversion and negative valence. The most frequently used words to describe transformation are: *create* (used 11 times), *new* (10), *development* (8), *reveal* (6), *construct* (6), and *regeneration* (6). Only then do we get to the first word, *reduce* (used 5 times) that might be conflated with loss, or have a negative valence.

Natalie had confirmed that Orff inverts, but we still hadn't resolved our debate about whether it was intentional or not. So, Natalie emailed her to ask. Orff's coy and genius response didn't solve the debate, but we can't go wrong by adopting it:

"I often think of our projects as starting with what is there plus a unit of transformation."

6

Scaling Subtraction

Using Less to Change the System

1.

It seems campy now, but when it was released in 1985, the music video for "Sun City" shone a serious spotlight on systemic racism. The song opens with a solo from the jazz trumpeter Miles Davis, then cuts to a pale woman in a red bikini lounging and smoking by a gigantic pool, and then it's over to rappers Run-DMC, in their hallmark fedoras, and underneath railroad tracks somewhere, for the first line: "We're rockers and rappers united and strong."

To begin the second verse, Pat Benatar fades in, dressed in black and with headphones on, apparently in a studio somewhere: "Separation of families I can't understand."

Soon there is an appearance by George Clinton, Mr. P-Funk

himself, who is sporting red and yellow hair extensions: "Our government tells us we're doing all we can."

Impossibly, Jimmy Cliff and Daryl Hall sing in perfect unison, even though Cliff is on set and Hall is bopping through, like a ghost, on a superimposed faded background: "Meanwhile people are dying and giving up hope."

Then, it's back to a streetscape for Duke Bootee, Grandmaster Melle Mel, and Afrika Bambaataa's shared "Sun City" moment: "You can't buy me I don't care what you pay."

Finally, building to a climax, we see: Bonnie Raitt: "It's time to accept our responsibility"; Bobby Womack: "Why are we always on the wrong side"; Bob Dylan and Jackson Browne, together, from what looks to me like an affluent subdivision in California: "Relocation to phony homelands"; and—for the last word—Bono (with a goatee): "We're stabbing our brothers and sisters in the back."

While some of these artists may regret their 1980s style, none will regret their participation in the project. Sun City was the name of a resort in Bophuthatswana, which the apartheid South African government claimed was an independent "homeland" for black people. By the time of the music video, most artists had stopped playing South Africa, but the resort in Bophuthatswana was a loophole by which acts could be recruited to play for an apartheid audience. They weren't playing South Africa, they were playing Sun City.

"Sun City" was part of the same public awakening as Leo Robinson's refusal to unload the *Nedlloyd Kimberley*. When the dockworkers left the cargo ship stranded and exposed, it became undeniable, physical evidence of a system of economic support for a racist regime across the world. By calling out the Sun City sham, the musicians were telling listeners why they, too, should stop propping up the racist regime.

For the world to help subtract systemic racism, they needed to see it. South Africans had been resisting since the apartheid policy was put in place in 1948. The racism became harder for outsiders to deny after the Sharpeville massacre in 1960, when police opened fire on peaceful protestors, killing 69 and injuring 180 more. The United Nations recommended cutting off military aid to South Africa, as part of what would be a long process of isolating the racist state. And yet, twenty-five years after Sharpeville, apartheid was still in place. As we learned in the introduction, trillions of dollars of economic divestment would eventually deal a final (subtractive) blow to apartheid. But for that to happen, others had to see the same underlying systems that Leo Robinson and the Sun City artists were trying to unmask.

Dismantling racism is a prime example of the systemic subtraction we will focus on in this chapter. In fact, as the history professor Ibram Kendi notes in *How to Be an Antiracist*, the phrase "'systemic racism' is redundant: racism itself is institutional, structural, and systemic." To subtract racism is to change the system.

A challenge when working in systems is that it can be hard to see the structures propping them up, whether racist or otherwise. In Kendi's telling, "We are particularly poor at seeing the policies lurking behind the struggles of people." To subtract the Embarcadero Freeway is hard enough, and the case for doing so is right there in front of us, blocking the waterfront. When it comes to removing barriers to fairness, just finding the redlined neighborhoods can require investigative journalism.

Seeing systems is particularly important for subtraction. We can blindly add good things without fully understanding the connections between them. But before we can get rid of something bad, we need to see and acknowledge it. Our ancestors creating the first cities, institutions, and policies didn't have to concern themselves with existing ones. But we do. That's why Kendi emphasizes that "the only way to undo racism is to consistently identify and describe it." Speak up, and often. This key lesson in anti-racism is relevant even if you don't have the platform of Run-DMC or Bono and even if the bigotry and corruption you encounter is not as blatant as apartheid. Because to subtract to improve a system, we need to see it first.

Racism isn't the only problem that becomes clearer when we identify and describe the system. Aristotle, for instance, had taught that stones fell to the ground when dropped because both the stones and the ground were made from similar materials. Smoke rose because it was more like air. But Aristotle's object-focused approach struggled to explain anomalies, like why arrows flew so far before returning to the earth. Galileo introduced forces, like inertia and friction, which were outside of the arrow, and better explained its behavior. Galilean science was a shift toward viewing objects as part of dynamic systems.

In general, systems are made of things and ideas; connections between them; and a surrounding field. That should all sound familiar. It is how we have described the situations we can change by adding and subtracting. Systems earn the "complex" tag when their behavior is unpredictable because of the dynamic interactions within them.

Most of the situations we encountered in high school physics and math were not presented as complex systems. We could predict the behavior of a swinging pendulum by knowing the height from which it was released. No matter how many operations in the math problem, there was a single correct answer.

Most of the situations we encounter in life are complex, whether a ship carrying South African wine, a concrete-covered downtown, or an overstuffed army training program. Simply introducing a human is enough to make a system unpredictable.

The subtracting insights we have already derived remain fundamental as we embrace this uncertainty. But more inherently complicated situations also deserve special focus, not just because they introduce more variables but also because they introduce new power to change. When lots of situations are linked together, when the outputs of one lead to the inputs of others, a small difference in our ability to see and follow through with subtraction can propagate. We can amplify our newfound subtracting talents by using them in systems.

A group of scholars in Germany recognized, in the early 1900s, that our understanding of human behavior could use a Galilean revolution of its own. Because one of their guiding questions was how humans acquire meaningful perceptions in such a chaotic and unpredictable world, this new lens on behavior would come to be known as the German word for perception: Gestalt.

The prevailing wisdom at the time was that discrete units of neurons, stimuli, and reflexes were the building blocks of behavior. As with Aristotle's object-focused approach, these units explained a lot about behavior. But the heretical scientists in Germany wondered if they could explain more by considering all these building blocks, and the systems they were part of, at the same time.

Just as object-independent concepts of inertia and friction helped Galileo connect experimental observations with causal influences, the Gestalt scholars sought the invisible forces that might help explain human behavior. Kurt Lewin's "valences" are one such force: the intrinsic goodness or badness of an object (or word, as Kate Orff knows). Lewin's "fields," meanwhile—which we have already borrowed to admire Savannah's squares and Harlem's pocket parks—grew to represent the sum of these forces: all the factors that influence behavior at a given time.

Fields and forces allowed Lewin and other social scientists to represent human behavior as a system of needs moving toward a goal. A force-field account of subtraction neglect, for example, includes the basic oversight my team found in our experiments.

It also considers adding instincts, like sense for quantity and desire to show fitness. It accounts for economic and other financial incentives, potential emotions from conflating less with loss, and so on.

Just as inertia and friction made Galileo's models more accurate, Lewin's approach better explained human behavior. Of course, when you consider all these forces together, you leave the comfort of predictability.

2.

Thankfully, Kurt Lewin was less interested in predicting small things than he was in improving big ones. Born into a Jewish family in Poland, Lewin moved to Germany for education, and had to leave Germany for the United States when Hitler came to power. Motivated by his interest in social problems, Lewin learned that one of the best ways to understand unpredictable systems was to try to change them. And he found that changing the invisible forces was often the best way to do so.

One way to transform a system is to add new forces that are working toward our goal. Such forces were added when Leo Robinson and the dockworkers sent food and medical supplies to anti-apartheid groups, or when the "Sun City" artists donated their proceeds.

Invisible forces can also be against us, though, in which case improvement comes from subtracting them. We can divest from apartheid. We can speak up to dismantle racism.

Like many great insights, Lewin's advice to subtract forces preventing progress is obvious in hindsight. Yet we've now seen over and over why removing obstacles is not likely to be our first thought. The same neglect that keeps us from subtracting in Legos and grids also applies to how we think about changing forces in systems.

My team confirmed this in our studies. Our university's latest strategic planning effort began, as most do, by soliciting ideas from students, faculty, staff, community members, alumni, and (of course) donors. All offered visions for transforming the complex system that is a university.

Gabe got her hands on the data, and as expected, adding was rampant. People wanted more study abroad grants, more mental health services tailored to international students, more housing options, and a new ice arena. Most of these changes seemed like progress to me (I didn't know there was a hockey team). But surely untapped potential remained. Out of about 750 ideas for change, fewer than 10 percent suggested subtracting.

When we overlook an entire category of options for a single change, that is bad enough. When it comes to systems, subtraction neglect is even more damaging—because, as it turns out, the options we are missing are better almost by definition.

Daniel Kahneman put it this way: "Lewin's insight was that if you want to achieve change in behavior, there is one good way to do it and one bad way. The good way is by diminishing restraining forces, not by increasing the driving forces." Lewin's "bad way" was to add—whether incentives for good behavior or punishments for bad behavior—because this increases tension in the system. Promising Ezra a cookie if he reads a book instead of watching a show increases his motivation not to swipe on his iPad. But promising Ezra a cookie does not make it any easier for him to resist the iPad. In fact, promising Ezra a cookie can make him more frustrated if he succumbs to the iPad temptation.

I can pursue the same goal, getting Ezra to engage with a book instead of stare at a screen, by removing the tempting iPad from the situation, placing it out of sight, or accidently letting it run out of batteries overnight. This is an example of Lewin's "good" way of changing a system, because it actually relieves tension.

It's the same with divesting from apartheid. All else being equal, adding incentives for anti-apartheid groups is not as good as removing the incentives propping up the racist system. To be clear, it's add *and* subtract here, as both changes move the system toward a less racist goal. But whereas the former gives the freedom fighters more ammunition for the struggle, the latter relieves the struggle itself. The very nature of systems—their size and complexity, and those all-important invisible forces—makes subtracting even more powerful.

Kurt Lewin was not the only scholar (or Kurt) from the Gestalt school to gift us subtractive wisdom. Kurt Koffka, in between being married four times to the same two women, originated the cliché about high-performing systems: "The whole is greater than the sum of its parts."

Koffka had found an uncomplex way to say that, no matter how much we learn about parts of a complex system, we still can't predict its behavior. The platitude above, however, while favored by sports announcers and motivational speakers, turns out to be a subtraction-blind mistranslation of what Koffka actually wrote. His original—and more accurate—wisdom was:

"The whole is *something else* than the sum of the parts."

Koffka was miffed by the "is more than" misinterpretation. He knew that the whole can also be less than the sum of the parts. As he repeatedly clarified, to no avail: "This is not a principle of addition."

Subtracting a part to enhance the overall performance of a complex system remains counterintuitive. That's why it was a surprise when traffic didn't worsen after the Embarcadero Freeway was removed. It was slightly less surprising when traffic got better after the 2005 removal of the Cheonggye Freeway, in Seoul, South Korea. By the time New York City closed Broadway to traffic at Times and Herald Squares, in-the-know planners had guessed their new pedestrian space might also come with reduced traffic on surrounding streets.

The road examples are the most concrete manifestations of a phenomenon proven by the German mathematician Dietrich Braess, who calculated that adding extra capacity to a system can sometimes reduce overall performance, or as Koffka would have recognized, that the whole is something else than the sum of its parts. In road systems, Braess's math plays out because whether one route is preferable to another depends not only on the capacity of the road but also on the density of the traffic, which depends on the complex behavior of human drivers.

When a new road is introduced, or when a freeway is subtracted, drivers try to optimize, until they think that other drivers have become set in their routes, at which point everyone stops optimizing. The new satisficed equilibrium may represent an increase or decrease in overall performance. Traffic in Seoul got better not because the system went from one optimal situation to another but because the road removal shook people out of one suboptimal situation into another one, which happened to be better. The removal could have made it worse too, but it's basically a roll of the dice, and it is most certainly not a principle of addition.

Braess's and Koffka's remove-may-improve wisdom is not limited to roads and traffic. It's been found in electrical power grids, biological systems, and even in my senior season of college soccer.

Our soccer team was a system in dire need of change. We had most of the same players returning from the team that had won the league the year before, and we had an extra year of experience. Yet we had barely qualified for the postseason as the fourth team in a four-team conference tournament. If we were going to win that tournament (which was our World Cup), we needed to beat the top-seeded team on their home field, where they had beaten us earlier in the season. Then, we would have to play the winner of a game between two other teams, each of which had recently beaten us on our home field.

Over the course of the season, our coaches tried everything to shake us out of our funk. They moved players to different positions. They replaced complacent seniors with upstart first-years. They yelled and coddled, dictated and listened. Now we had just a few practices to prepare for the conference tournament, we desperately needed to perform better, and our coaches had exhausted seemingly every option for getting us to do so.

As you may have guessed, our coaches' epiphany in those final few practices was to subtract. During those practices, we played with nine players, two fewer than the eleven we were allowed in games. That shook us out of our suboptimal equilibrium. Only after we began to function as a system of nine did the coaches add two players back on. We were a transformed system. We played better than we had all year (and, as you also may have guessed, we won the tournament).

We don't think to remove the Legos, or extra words, or even squares from Andy's grids, and we don't think to remove things from a malfunctioning system. Doing so is so unimaginable that even Kurt Koffka couldn't get the idea across. When the fact that subtracting can improve a system was proven with basic math, the finding became known as "Braess's paradox," as though it was an anomaly beyond understanding.

3.

After soccer was no longer the only system that interested me, and before I became the kind of doctor who is useless in an emergency room, I managed the design and construction of large buildings. That is when I formed my opinion that calling systems "complex," while technically accurate, draws our focus away from where it should be.

A typical project I worked on was a school in Elizabeth, New Jersey, a city just south of Newark. This school, which now serves a few hundred pre-K through eighth graders, was part of a statewide program to provide sorely needed upgrades in long-neglected communities. The school building is three stories tall and built over an underground parking garage. There are a few dozen classrooms, a gymnasium, and a cafeteria. Outside there is a playground, paved lanes for parent and bus drop-offs, and security gates and fencing.

I like to think that I helped build schools. My brother likes to remind me that I didn't actually "do" anything. These differing perspectives are because my job and my company existed not to physically build the school but to deal with the systemic complexity of doing so.

To create this straightforward school in Elizabeth, hundreds of materials had to be chosen, procured, and assembled. Even for something as basic as a toilet handle, architects and engineers had to specify how the material should look and perform. Someone like me, except working for the contractor, would review these specifications and send them to suppliers of toilet handles. Interested suppliers would return to the contractor reams of information about what they deemed their most worthy toilet handles. This information could be hundreds of pages long, with glossy photos, written descriptions, calculations, schematic drawings, prices, and verifications from independent testing agencies of how the supplier's toilet handle did everything the specifications required. The contractor then evaluated the toilet handle candidates and forwarded their favorite to the designer. Sometimes, a subcontractor or two would do their own review and approval between the supplier and the primary contractor. An architect (to make sure the handle looked right) and an engineer (to make sure it would get the job done) reviewed the options. If, for example, the toilet handles were for a teachers' bathroom, a sage architect might head off future strife by seeking approval from someone who spoke for the teachers. After all of this, the initial submission might be approved by the architect, or they might ask the contractor for more information, or tell them to try again.

We already are inundated with a system of activities, information, and relationships, and all we have are toilet handles. This process had to be repeated thousands of times for all the materials that went into the school.

One response to the complexity of material acquisition is to hire some recent college graduate to keep track of it all. So I went through the architects' specifications and made a list of all the required materials. As the project progressed, I kept track of which ones had been requested from suppliers, which had been sent to the architects, which had been approved, and so on. It's a perfect task to motivate the recent grad to strive for promotion or to consider graduate school.

There is complexity in choosing and acquiring appropriate materials, and there is complexity in putting them together in a safe and timely manner. To build that school, thousands of humans were involved in thousands of ways and at thousands of times. Architects, engineers, and contractors, all with different specialties and motivations, and also the students, teachers, and staff who would use the building, the custodians and engineers who would maintain and operate it, and the administrators and state officials who approved the plans and signed the checks. If that wasn't enough, our single three-story building was, like most projects, part of a much larger school construction program. Someone had to consider budgets and schedules and how they aligned with other projects, funding commitments, and political happenings, like whether the school principal who pushed hard for the chrome toilet handles was likely to be reappointed. Every person I dealt with knew more than I did about some aspect of the project. I was supposed to help them see how they fit into the system.

Now that you know why I am jaded by devoting thousands of hours to representing complexity, we can move on to looking for essence. Essence is the soul complexity, its irreducible building blocks. All the complexity brought by biological evolution, for example, builds from the genetic code within DNA, which is represented by just four letters, combined into patterns of three letters each. Genetic code is essence.

Critically, finding essence in a system is not a matter of categorization, as when I kept our huge cache of construction submittals broken into categories by function. (The toilet handle is in section 224000 of the MasterFormat specification system.) Segmenting did keep details retrievable and, by breaking the project into smaller pieces, made it more comprehensible.

My ultimate job, though, was to understand how material availability was affecting the progress of construction—and for that, I couldn't just segment out the plumbing fixtures. I needed to look at all the categories, and when I combined the smaller lists, I was right back to where I started: with too much information. The more we broke our system up into discrete steps, the more we lost those vital connections between the steps. Too many subdivisions of labor led to a loss of connection between them, and more work for me. I had simply segmented what we had; that didn't tell me what it all meant.

There were countless steps, relationships, and reinforcement that made up the building process. And it was detrimental to consider all of them when making decisions.

The construction superintendent did far more than I did to make that school in Elizabeth, New Jersey, a reality (and was paid accordingly). He never looked at my list of submittals or the full construction schedule. Instead, he had his short lists, sometimes scribbled on spare paper in the pockets of his well-worn jeans, and often only in his mind. It had taken the superintendent decades to develop these far more valuable lists of just the few materials and steps most critical to the project's progress. What I had not learned in any of my college engineering classes (or soccer practices) was that, as decision aids for complex systems, abridged lists work better than comprehensive ones.

To understand why lists with less win, we need to understand working memory. This is the cognitive system that temporarily holds the information we have available for processing. In other words, working memory holds the ideas we can quickly bring to bear on our change efforts, whether the situation is an unbuilt school, a grid pattern, or a vessel loaded with cargo from South Africa.

We cannot both represent every detail of a system and also expect to use the information. This tension between detail and usefulness has long provided fodder for the imagination. The writers Lewis Carroll and Jorge Luis Borges have each fantasized about the quest to make a perfect map. Driven to include the entire complex system that the maps are meant to represent, the imaginary mapmakers keep adding detail. Eventually, the maps perfectly describe their respective empires—and are just as big. In Borges's telling, the map ends up blocking out the sun.

These writers used stories to explore the limits of detail. The psychologist George Miller used experiments to show that our working memory hits limits long before we've built a full-scale map.

"My problem is that I have been persecuted by an integer," is a delightfully unlikely opening line for an academic paper, especially one from 1956. It is how Miller begins his treatise on the limits to working memory, in which he reveals the offending integer to be seven (plus or minus two). Miller's paper is often interpreted to argue that seven is the exact number of things we can simultaneously think about and use. Let's not fixate on that number, because our understanding of working memory has advanced since Miller, as he hoped it would. But his finding that our working memory has a severely restricted capacity, often *below* seven things, has been shown repeatedly.

What this means for us is that, when we are considering complex systems, we need to avoid overwhelming the capacity of our working memory. Thousand-item submittal lists and perfect recall of South Africa's legal codes may be required for lawsuits, but they are full-scale maps blocking the sun when we need to both remember and use information. To transform systems, we need to find the essence, which means we need to subtract detail.

The question is: What to leave in and what to take out?

Having read all that I can about this rapidly evolving science of systems, Dana Meadows remains my first-stop authority. Meadows pioneered the study of complex systems as part of a team at the Massachusetts Institute of Technology. Like the perfectionist mapmakers, Meadows's team built detailed models of the world, overlaying earth systems with human variables like economic production, pollution, and use of nonrenewable resources. Meadows also sought and shared essence, doing so over three decades of teaching and writing about complex systems while at Dartmouth College. She knew complex systems so well that she could explain them simply. Her timeless book *Thinking in Systems* is proof.

Thinking in Systems emphasizes finding the goals of the system. As Meadows put it, we discover these goals by asking, "What is the system trying to achieve?"

The question is simple, but we often either forget to ask it altogether, or we pay it so little attention that we end up assuming an incorrect goal. I never thought much about the goal of tracking the submittal process—to know how material status was affecting overall project progress. If I had analyzed the goal, I may not have been satisfied with collating the status of every last material and instead may have come closer to the construction superintendent's wisdom. The scraps of paper in his pockets were knowledge we could actually use. With my own priority lists, I could have evaluated whether I should spend my finite time and social capital cajoling the architect to approve the toilet handles, or whether I could leave her alone to work on the structural steel.

I'm not suggesting that it is easy to extract essence from detail. If our animal desire to show competence kicks in for file folders and Harvard essays, then surely it will as we try to show competence building schools.

But if the challenge to subtract is greater the more complex the system, then so is the payoff. Subtracting unnecessary detail is how we clarify places and ways to intervene. We strip away the particulars to see that what matters first is that apartheid is understood, and we realize that an idle ship or a star-studded protest anthem will help do the trick. We realize that identifying and describing racism is the first step in undoing it. Now, we are gaining power to change the entire system.

4.

As an emergency room doctor, my sister, Carrie, constantly deals with complex systems. Whether a choking toddler, a diabetic low on insulin, or a grandparent with hip pain, every new patient Carrie sees is a unique and unpredictable case. Fortunately for them, she spent years in medical school and uses beach vacations to build and refine her vast knowledge of how to help patients.

With each new case, Carrie not only needs to access medical expertise from her sleep-deprived brain, she needs to apply it to the patient at hand. The same affliction in different patients can demand an entirely different course of action. A teenager who is unresponsive after their first encounter with alcohol needs different help than the passed-out fifty-year-old who is brought in a several times a year.

And rarely is Carrie dealing with a single patient. No matter how long you have been waiting to have your cut finger stitched up, when the man who just had a heart attack is wheeled in, you are going to be waiting longer (and probably with a new perspective on your own situation). My sister needs to consider each individual's condition and also multiple patient conditions relative to each other.

Carrie holds all these situations in context—she needs to see the field. When only one anesthesiologist is working and two people require emergency surgery, she has to decide which surgery is more urgent, and which can wait until the on-call anesthesiologist drives in. I bet my sister even knows where the on-call anesthesiologist lives and what the traffic is like between there and the hospital.

I could go on about my impressive sister, but you get the point. She faces vast and unpredictable systems, with infinite change possibilities, and not enough time to consider them all, because unseen patients are deteriorating. Given the number of variables, and Carrie's hard-earned ability to navigate them, I was surprised to learn about the simple triage process she uses to do a first pass at who gets what care. The process, used by highly trained emergency room doctors everywhere, guides them through some version of the following sequence:

Does this patient require immediate lifesaving intervention? (Is this a patient who shouldn't wait?)
How many resources will this patient need?
What are the patient's vital signs?

That's it. The triage process strips down emergency room systems to their essence, which allows my sister to make improvements.

Carrie went to medical school at the Johns Hopkins University, where she took a class with a young doctor named Peter

Pronovost, who demonstrated how subtracting detail could save lives. Pronovost wanted to improve the practice of inserting central line catheters, which are those thin plastic tubes used to draw blood or administer fluids and medication.

Optimizing catheter insertion does not sound as heroic as transplanting a heart or precisely extracting a brain tumor. None of the medical dramas my sister watched as a kid built their story lines around catheter placement. But Pronovost knew that infections from these catheters were causing about thirty thousand deaths each year in the United States—killing roughly as many people as car accidents.

To insert a catheter line, there are dozens of steps, each requiring a mix of thought, judgment, and physical skill: anesthesia is administered; the catheter must be placed to prevent air embolisms; placement is confirmed by x-ray; and so on. Inserting a catheter on a dehydrated ten-year-old is also much different from inserting one for a concussed offensive lineman. A *summary* of the guidelines for inserting central catheters runs thirty-five dense pages in a medical journal. Those pages synthesize and are supported by thousands more, from other journal articles, expert opinion, and clinical data.

To prevent infections, Pronovost and his team considered all this complexity and more. Then they proposed that medical professionals: wash their hands with soap; clean the patient's skin with antiseptic; put sterile drapes over the entire patient; wear a sterile mask, hat, gown, and gloves; and, put a sterile dressing over the catheter site.

The steps are simple. The results are striking. Reminding doctors to follow these steps has almost entirely eradicated catheter infections at the Johns Hopkins University Hospital, and in early-adopter states, including Michigan and Rhode Island. Subtracting detail—to get to the essence of the system—has saved thousands of lives.

Of course, we cannot bypass the detail altogether. Peter Pronovost and his team knew that there were more than a hundred surgical fires each year and that, sometimes, people died from these fires. They did not include limiting fire risks in their catheter checklists because, though adding more checks may have eliminated a fire or two, this would have made the list longer and therefore less useful to protect against the greater infection risks.

Likewise, I cannot use the emergency room triage process. That would be a deadly version of lazy less. But Carrie earned all As in college, spent four years in medical school, passed through a gauntlet of standardized exams, and spent three more years as a resident learning and doing alongside experienced emergency room doctors. She's now spent most of her working life in emergency rooms, and much of her nonworking life thinking about them.

Given all that, Carrie's skills are best directed through simple steps. The care you get—and, yes, how long you sit in the waiting room—still depend on her hard-earned expertise and innate understanding. But by subtracting detail, the triage process helps emergency room doctors focus on the essence of their tasks. With their refined understanding, they can move on to asking whether subtracting might improve the system itself.

5.

A checklist for subtraction is in order at this point. We can keep it in our working memory as we move forward—from seeing how subtraction works in systems to using it to transform them. Among the takeaways we are collecting to help us find and share less, this checklist can help us remember the essential steps of doing so.

Subtracting detail *before* trying to change the system, like emergency room triage, will come first on the list. Persisting to noticeable less, like Springsteen's "samurai record," is on there too. The other two steps are subtracting first and reusing your subtractions. We can quickly learn and remember them with the help of Jenga and doughnut holes, respectively.

The game Jenga is so low-tech it might seem to have been around forever, like chess or cards. In fact, Jenga originated in the mid-1980s, brought from Ghana to Britain and then to the rest of the world by Leslie Scott, who has gone on to become a serial toy inventor. That's right, in the history of toy development, humanity thought to subtract from building blocks about the same time as it produced the computer game Tetris.

To set up Jenga, blocks are prearranged in a stable and solid tower, with three blocks to a row, each row perpendicular to the one beneath. Players take turns removing one block at a time and replacing it on the topmost level of the tower. As the tower grows taller and less stable, it becomes harder to complete your turn without knocking over the tower. The loser in Jenga is the one who does so. Everyone else feels like a winner.

Legos and Jenga both satisfy Ezra's desire to build. It is the different rules that lead to entirely different outcomes. In Legos, Ezra first assembles them as prescribed in the set's instructions. Then, he plays with the assembled version for up to an hour. Whatever he built then stays on the floor of our addition until the next time the room becomes impassable, which is typically within a week. At this point, Ezra decides whether his most recent creation warrants indefinite display on a windowsill, or whether it can be tossed into one of the drawers under his Lego table, to be scavenged for future freestyle building. By this time, he has begun making his case for buying the next set of Legos.

Legos encourage endless adding, especially when you have a dad who supports your habit.

In Jenga, the rules promote balance. Jenga forces us to subtract first, requiring that we pull out a block from one of the lower levels before we add to the top level. Sure, Lego's adding approach has been good for business; but so has Jenga's mandate to subtract first. It was the game's novel subtracting rules that Leslie Scott copyrighted, to the tune of one hundred million copies sold.

It's not just toys telling us that subtracting first can amplify the power of our changes. Project management textbooks remind students and their professors that, when there is a series of changes in which outputs of early ones become inputs to later ones, the early changes tend to be more influential and cost less to make. Catching a flawed toilet handle when it is a drawing is better than finding out after it is installed. Washing hands to keep the catheter site clean saves more lives and costs less than treating infections after the fact. In this same way, subtracting first diverts us from the well-trodden path to more.

So after you have subtracted detail to find the essence of the system you wish to change, consider subtracting first, as in Jenga. Then persist to noticeable less. Last but not least, don't forget that you can reuse your subtractions.

Doughnut holes provide a memorable illustration of this step in the subtracting process. As with the transition from solid blocks to Anna Keichline's K-brick, it took a long time for someone to realize that fried dough could be improved by removing from it. The most well-documented story dates the innovation to 1847 and credits a teenager in Maine named Hanson Gregory. Young Hanson asked his mom why her fried cakes were always so soggy in the middle. She said she didn't know. So the teenager took out a fork and punched a hole in couple of the uncooked rounds of dough. His mom fried them. Doughnuts finally had holes.

Removing a ball of dough from the center of the doughnut lets it cook more evenly—and provides more surface area for cinnamon sugar. Less is literally more. Not surprisingly, the post-Gregory years have been good for doughnuts. At the 1934 World's Fair, doughnuts were declared the "food hit of the century of progress." Around this same time, a bakery in New

York City grew into the first doughnut chain, Mayflower Donuts, which adapted the optimist's creed for their boxes: "As you ramble on through life Brother, whatever be your goal, keep your eye upon the donut, and not upon the hole." For a long time after Hanson Gregory's innovation, that's what people did. They kept their eyes on the doughnut.

It would take more than a century for the holes to turn from functional void to salable solid. As we now know, those little bits of subtracted dough have plenty of appeal on their own. Whether you prefer Dunkin' Donuts Munchkins (1972) or Tim Hortons Timbits (1976), reusing the subtraction has made for another stream of income.

Reusing our subtractions allows us to exploit an advantage of subtracting. When we add to change a system, we are left with the improved system. But when we subtract to improve a system, we are left with the new-and-improved system, *plus* whatever we have taken away from the old one. What is true in doughnuts is true in consequential changes. When the state of California subtracted $11 billion out of apartheid South Africa, that was $11 billion they could invest elsewhere. Just because a subtracted bit was holding back one system, that doesn't mean it can't be useful somewhere else.

Like emergency room doctors, we now have a checklist that gives us room to act and adapt.

- Subtract before improving (e.g., triage)
- Make subtracting first (e.g., Jenga)

- Persist to noticeable less (e.g., Springsteen's *Darkness*)
- Reuse your subtractions (e.g., doughnut holes)

These four steps can direct our expertise. We can keep the steps in our working memory as we go to work. Let's call them the *lesslist*.

You will notice that these four items do not embody all the takeaways from this chapter: the "Sun City" artists' reminder that we need to see systems to subtract from them; Lewin's wisdom that removing barriers is the "good" way to change systems; and Koffka's insistence that transforming systems is "not a principle of addition." Nor do these four items summarize the first six chapters. Just as my sister brings her hard-earned expertise to emergency room triage, we need to bring our newfound subtracting skills to the lesslist.

So let's practice using these new tools. As we're about to see, it's urgent. Because human behavior, including subtraction neglect, has become so powerful that it is changing the complex system that supports all life: Earth.

7

A Legacy of Less

Subtracting in the Anthropocene

1.

Our new geological epoch, the Anthropocene, is defined by the unprecedented fact that a single species (ours) has become the dominant influence on our planet's well-being. We are the biggest force in the ultimate field.

The Earth's environment is an incredibly complex system, which means we need to get to the essence. Let's start with a masterful synopsis that has stood the test of time and checks in at around two thousand words.

For those of you who are behind on your kids' book reading and haven't watched the movie version, here's what you need to know about Dr. Seuss's *The Lorax.* An enterprising young man, the Once-ler, comes across a valley of thriving animals and abundant Truffula Trees, their foliage swirling in bright pastel colors. The Once-ler chops down one of these trees, removes the silky foliage, and knits it into a garment, which he brands a Thneed. Enter the title character, the beaver-bodied and walrus-mustached Lorax, whose job is to "speak for the trees." The Lorax scolds the Once-ler for axing the Truffula; the Once-ler points out that there is an entire forest of them. The story proceeds until the Lorax's worst fears are realized and the Once-ler learns his lesson. Thneeds become a fashion trend; the Once-ler scales up his operation. He builds a factory and cuts down more and more trees, faster and faster. What had been a vibrant valley becomes a polluted wasteland. The Lorax lifts himself up by his tail and flies away. With no Truffulas left, the Once-ler can no longer make Thneeds. He shutters his factory and spends the rest of his life holed up in a Lerkim on top of his store, reflecting on what went wrong and how he might fix it.

Seuss's playful proof of how people depend on the environment has forced me into some parenting I did not expect to have to do so soon. As a four-year-old, Ezra worried that his clothes came from a factory like the Once-ler's. In *The Lorax,* the Once-ler's factory produced "such smogulous smoke" that the Swomee-Swans could no longer sing, and the factory dumped "gluppity-glup" into the water where the Humming-Fish lived. I assured Ezra that, since *The Lorax* was published, we have figured out ways to make clothes without shooting as much swan-choking smog into the air, or dumping as much waste directly into rivers. I proudly reported that some innovative members of his species—including Ray Anderson, the late CEO of Interface carpets—have improved their factories so that the water leaving it has less pollution than the water that enters.

The situation now, I ashamedly explain, is that human activity is affecting our environment on a much larger scale. I tell Ezra that humans have burned enough fossil fuels to disturb even the places where there are no Thneed factories.

I don't want my son to lack respect for grown-ups yet, so I clarify that the energy these fuels unleash has, like the human energy released by agriculture, allowed far more people to lead far better lives than ever in history. And I clarify that, for much of the time we enjoyed the benefits of fossil fuels, we did not know the carbon dioxide they release was stressing our environment on a planetary scale. But now we do know.

While more greenhouse gases in the atmosphere may not immediately silence Swomee-Swans, I tell Ezra, the overload is changing the environment faster than many species can adapt. For four hundred thousand years, the atmospheric concentration of carbon dioxide hovered between 180 and 280 parts per million. When Truman gave his anaphora of more speech, roughly a century after humans had begun to harness the power of fossil fuels, atmospheric concentration had risen to 310 parts per million. This was out of the historical range but safely below the 350 parts per million that scientists now believe can help us avoid the most destructive changes. Then came seventy years of better equals more, fueled by human burning of fossil fuels. When I talked with Ezra, the concentration of carbon dioxide in our atmosphere was 413 parts per million—and rising.

My son is tuning out the history lesson, even the number-free version, so I go back to how people are harmed by climate change. I tell him that change is not the problem. Species have always moved about the earth to find environments conducive to their lives. The Lorax flew beyond the smog and out of the Truffula valley. My ancestors migrated south to avoid cold and find mammoths. Maybe life was better where they ended up. There's nothing wrong with a little moving.

What's different now, I say, is that we're not just smogging up a Truffula valley, we're changing the entire planet. And we're doing so at a rate faster than we—and the systems that we rely on—can adapt. I tell Ezra that glaciers are melting, storms and droughts are harsher, and that islands and coastal cities are being battered and even overtaken by rising seas. I'm almost back into full-on professor mode when he interrupts me:

"I know already."

Turns out, he had learned about climate change at preschool.

I wonder whether Ezra's preschool teachers have told him about the World Health Organization estimates. As Ezra grows up, climate change will cause about 250,000 deaths every year from heat stress, malnutrition, and malaria alone. I hope they haven't told Ezra that a quarter million human deaths per year is a conservative estimate, not including those linked to ancillary effects of climate change like reduced farming yields and population displacement.

Those quarter million deaths per year would not include Alan Kurdi, the three-year-old whose image reminded me of a napping Ezra, facedown, toddler sneakers up. Except, Kurdi wasn't napping. He was photographed after washing up on the beach, having drowned in the Mediterranean Sea. As he had throughout his short life, Kurdi was migrating because of the Syrian civil war, a social unraveling likely hastened by a drought made worse by climate change.

I withheld those statistics and Kurdi's story, leaving Dr. Seuss to teach my son why the environment and humans are indivisible. As Seuss explains, when the Once-ler clears the Truffula Trees, he sets off a chain reaction that makes the entire valley uninhabitable. Not just for the Swomee-Swans and Brown Bar-ba-loots but for the Once-ler's friends and family as well.

The Once-ler was the dominant force in the Truffula valley system. Now, Ezra and his fellow humans are the dominant force on our planet. He plans to share the official term, Anthropocene, with his classmates.

My son seems sad to learn that he is inheriting this situation, so I end by reframing the Anthropocene in a more positive light: Ezra is part of the first species with the power to improve the planet. The question, as the Once-ler eventually recognizes, is how we will direct that power.

As Dr. Seuss was putting connections between humans and the environment into verse, Dana Meadows was building the computer model that would allow her to study these connections. Meadows and her team at the Massachusetts Institute of

Technology collected and modeled all the relevant information they could find about the trajectory of human life on earth and the environment that supports it.

In 1972, Meadows and her team shared their findings in a report that became one of the most influential nonfiction books of all time, selling, by some estimates, more than thirty million copies. The title distilled the researchers' conclusions to their essence: there are *Limits to Growth.*

The Once-ler had said to the Lorax, "I'm figgering on biggering and BIGGERING and BIGGERING and BIGGERING." *The Limits to Growth* called out this attitude as the crux of the problem. Adding has been good; we live better than royalty in previous generations. I have two children who wouldn't be here without modern medicine, a house for them to live in that's warm in the winter and cool in the summer, and the world's knowledge at my fingertips. Surely we should improve the human experience for as many people as possible, now and in the future, or as conservationists say, for "the greatest good of the greatest number for the longest time."

Assuming this desire to provide the most good for the most people remains, our approach to it must evolve. It's not enough to buy fewer Thneeds—and we cannot fly off beyond the smog like the Lorax.

So, what should we do?

This is a polarizing question, even among the well-meaning. Some Anthropocene improvers side with the Lorax. They emphasize limits. They appeal to the logic that infinite growth is impossible on a finite planet. They cite studies showing that our earth has a limited carrying capacity for human life. They point out that those 413 parts per million of carbon dioxide in the atmosphere exemplify how humans are exceeding the safe operating conditions for life on earth. Loraxes who like data may note that our rate of economic growth is matched almost exactly by our use of fossil fuels, which tracks almost exactly with the carbon dioxide we have added to the atmosphere. Given the direct relationship between gross domestic product and harmful emissions, these Loraxes might even contend that the only way to stop plundering the planet is to limit economic growth.

Others side with Seuss's Once-ler. To be clear, these new Once-lers are not climate change profiteers. These are not fossil-fuel billionaires and the media influencers they buy to sow lucrative (for them) distrust in climate science. These are not the corporations and the elected officials in their debt who conspire to keep exploiting the climate commons. As with racism, we need to constantly identify and describe these entrenched barriers to climate action, so that we can dismantle them. But these uncaring actors are not Once-lers as we will use them here. Seuss himself reveals that the original Once-ler never intended the devastation he wrought; he didn't take his profits and flee the wasteland he created. He has stayed there, in his Lerkim on top of his store, protecting the last Truffula seed. He has "worried about it, with all of [his] heart."

Like the new Loraxes, the new Once-lers bring good intentions and logical, science-informed arguments. Team Once-ler points to a history of scientific innovation and striving for progress. Humans have already found ways to do more with less, even extending carrying capacity beyond what some of the *Limits to Growth* models forecasted. Interface CEO Ray Anderson, with his water-cleansing factories, was a business Once-ler; the TED star and *Factfulness* author Hans Rosling, with his Tufte-esque graphics of global progress, was a development Once-ler. Lisa Jackson, in her work with the EPA, was a legislating Once-ler. All three respect, but don't get hung up on, limits to growth. They rationalize that we will keep improving and use the proceeds of doing so to figure out the resource constraints as we go.

Nearly half a century after Seuss published *The Lorax,* most thoughtful people and groups are far on one side or the other of this Lorax to Once-ler continuum. But to improve our Anthropocene situation, we need the best of the Lorax and the Once-ler, which rarely happens.

Subtraction can help bridge the gulf. Sure, Loraxes are right that continuous progress on a finite planet breaks down when it relies on more and more finite resources like farmable land, water, and fossil fuels. But devastation need not be the end result—if progress comes through subtraction.

Kate Orff has Once-lerish multitudes, from her use of science and technology, to her perspective that change is positive, to her audacious belief that her ideas can change vast physical systems. Yet Orff's breakthrough in Lexington—getting rid of infrastructure that was already there—required those same

Once-ler selves be applied in a counterintuitive way. She recognized limits in things, like the Lorax, to create new possibilities.

It's not just Kate Orff. If we reflect on *Subtract*'s heroes, we find a balance of the thoughtful Lorax and ambitious Once-ler. Like the Once-ler encountering the Truffula valley, Sue Bierman came to San Francisco and created a waterfront. Elinor Ostrom was Lorax-like in her appreciation for the very real limits of common-pool resources, but she was a Once-ler in not ruling out, and then painstakingly confirming, the potential of human ingenuity to manage these commons.

Unfortunately, the likes of Orff, Bierman, and Ostrom are the exceptions. Subtraction is a chronically missed option, as we know. When it comes to the Anthropocene, this oversight becomes a fundamental flaw because less may be the key to more good for more people for more time.

2.

Let's apply our lesslist to the Anthropocene and see if we can glean some ways to leave a legacy of less.

Step one is pre-action subtraction. Triage helped in my sister's emergency room. Can we follow her lead when the patient is our planet?

The Anthropocene contains interwoven goals: economic ones like selling more doughnuts; social ones like dismantling racism; and environmental ones like staving off the catastrophic effects of climate change. All these goals affect one another. The effects of climate change, for example, fall disproportionately on economically poor minorities. Each goal alone has complexity beyond what any computer model can capture. The Intergovernmental Panel on Climate Change (IPCC) convenes hundreds of scientists, reviewing the work of thousands more, to provide *summaries* of the science on climate change. One of the IPCC's latest reports runs 167 pages, and those 167 densely packed pages are a synthesis of other syntheses. There is no fat to cut. A single line in an IPCC report could represent multiple lifetimes' worth of study or action.

Within the goal of maintaining a climate that supports life, there are also countless interdependent issues: changes in the atmosphere, seas, and extreme events; impacts on food and water systems and biodiversity; and projections and proposed paths forward through decision-making, finance, and policy. Each of those sub-issues could have its own report.

One response to a 167-page report that is just the tip of the iceberg is to reassure ourselves that every little bit counts. As long as we do something to improve the situation, it matters. If we can coordinate with others, even better. I think this is nonsense.

We need to prioritize. If it's true for construction projects and catheter insertions, it absolutely has to be true for the Anthropocene. There simply are not enough human resources to devote to every possible response to climate change. We can't do it all. We need to do what will make the most progress.

Carolina Mauri has Once-ler ambition to make progress.

After competing as an Olympic swimmer, Mauri earned her law degree and has since been instrumental in setting Costa Rica's climate change policy, in multinational agreements, and in her home nation's government. Before trying to improve the system, Mauri took what you now know can be a hidden and hard first step. She subtracted information about the situation itself.

Remember, to subtract information, it needs to be there in the first place. Mauri considered complexity. As an expert on climate change law and policy, Mauri knows that adapting to environmental changes is as important as lessening them. She knows that Costa Rica is just one nation—one herder in the vast climate commons. And she knows that any climate goals had to also consider impacts on poverty reduction, health, and even economic growth.

Mauri considered all of that detail and more. Then she got rid of it to find essence.

No matter how accurate, exhaustive prescriptions can draw focus away from more influential ways to take Anthropocene action. Peter Pronovost didn't include any fire prevention measures on the catheter insertion checklist. Such measures would have saved a few lives but cost many more in distracting attention and resources away from much more harmful infection risks. Likewise, Carolina Mauri didn't want to waste her time and others' by elevating programs like recycling and composting to the same urgency as truly lifesaving measures like drastically slashing carbon dioxide emissions.

Mauri subtracted complexity—and then she subtracted

some more. In 2007, Costa Rica announced the essence. They would become the first carbon-neutral country by 2021, the two-hundredth anniversary of their independence from Spain. It's not an easy goal, but it's a useful one.

Triage fits in my sister's working memory, guiding her choices in the emergency room, just as the 2021 vision guided Costa Rica as they made more granular decisions about budgets, laws, and incentives. Using fossil fuels took Costa Rica away from their 2021 goal, so that activity was taxed. Energy sources that don't emit carbon, including solar, wind, and biofuels, were incentivized. Costa Rica may be a relatively small herder in the climate commons, but with pre-action subtraction, they set an example we can learn from as we take on our grand challenge.

Step two on our lesslist is to make subtracting first. Let's see what it might mean to play Jenga in the Anthropocene.

The doctrine "Reduce, reuse, recycle" arose from the same realizations about human impact on the environment that led to *Limits to Growth* and *The Lorax.* If we reduce—buying fewer Thneeds—new Truffula Trees might have time to grow before the Once-ler's "Super-Axe-Hacker" whacks them off "at one smacker." Likewise, by reusing old Thneeds, we could slow the Truffula slaughter. Recycling is last priority because, while recycling Thneeds would give Truffula foliage another use, it would also require resources like energy and water as part of the remanufacturing process. Reduce, reuse, recycle is triage for natural resources.

The three Rs suggest ways to stem the flow of emissions into the atmosphere. If we reduce consumption of fossil fuels, we reduce climate-changing greenhouse gases—of course. We need to do everything we can to stop adding emissions to the atmosphere. We also need to take emissions out, though, which means the three Rs are not enough.

The three Rs can in fact be harmful, by inducing the same set schema that limits the options of little kids (and Blaise Pascal) who think negative numbers are "unpossible" because they see zero as an unbreachable baseline. If we only use the three Rs, we treat the current level of emissions in the atmosphere as our unbreachable baseline. And as I did not have to explain to Ezra, our current emissions are already over what scientists have deemed a safe legacy.

When the current situation exceeds planetary boundaries, we need to subtract first. *Remove* must become the first R.

Once Costa Rica homed in on their goal of carbon neutrality, they looked at ways to get there. Obviously, Costa Rica needed to add fewer emissions to the atmosphere, like Ben ringing his no-bell to slow the rate at which he overcommits himself. But to have any chance at achieving carbon neutrality by 2021, they needed to avoid thinking of the carbon already in the atmosphere as an unbreachable baseline.

As Ezra also learned in preschool, trees pull carbon dioxide out of the atmosphere. Sure enough, restoring forests is one of the most cost-effective ways to improve the Anthropocene. Especially for nations like Costa Rica, which combine a perfect climate for growing lush forests with lots of open space for doing so (thanks to a history of human deforesting to draw the Lorax's ire). Now, we all need to do our own Anthropocene checklists. Reforestation may not be an appropriate change for your nation, city, or backyard, but it is for Costa Rica—and it's not to be overlooked when we subtract first and consider ways to "remove" emissions.

Thank you, Jenga.

While we are here, let's channel Kurt Lewin: removing emissions is the "good" way to engineer the climate. Climate engineering encompasses a growing host of ways to counter the effects of climate change by meddling with other complex and large-scale earth systems (the very systems we have thrown out of balance without even trying).

Some of the ideas: A fleet of airplanes could spray aerosols into the atmosphere to block sunlight from getting to the ground. Adding space mirrors would pursue a similar sunlight-blocking goal by reflecting the sunlight back into space. Other proposals would lighten the color of the earth's surface, to reflect more sunlight. Or dump tons of iron filings into the oceans, changing one unpredictable complex system in an attempt to improve another one. I hope you are noticing that these climate engineering proposals all add.

Kurt Lewin acknowledged that adding and subtracting can make desired behavior more likely. But only removing barriers promises the extra benefit of relieving tension in the entire system. Offering Ezra a cookie if he turns off his iPad without getting mad makes him less likely to get mad when the machine goes dark. If he does get mad, though, it's going to be the "no iPad" madness plus the "I'm not getting a cookie" madness.

Kurt Lewin was talking about social situations, but this same principle of tension applies in environmental systems. Spraying aerosols into the atmosphere is like adding the cookie incentive. It may reduce global warming—and it will definitely add more human-made stuff to our atmosphere, which will add tension and volatility for the next unforeseeable situation. In the systems that support human life, which we evolved to thrive in and then managed to disturb profoundly, the less volatile, the better.

Maybe climate engineering will help repair our legacy. To be fair, some plans do put subtracting first, proposing to remove carbon dioxide from the atmosphere. In my opinion, even the adding options should be on the table. But if we overlook less in our climate engineering, that would be a classic example of the folly of confronting a problem with the very mindset that created it.

Step three is to persist to noticeable less. As we have seen, subtraction has a publicity problem. Whether a brownstone in Harlem that has made way for a pocket park, or an unnecessary adjective culled by an editor, what has been subtracted is no longer visible.

The noticeability challenge is great in the Anthropocene. Not only is subtraction invisible, but the changes themselves, whether adding or subtracting, tend to be diffuse over space and time. Whereas our senses help us notice changes in weather, to notice changes in climate requires careful measurement over long periods of time, and then, ideally, distilling of these measurements into Tufte-esque graphics. No matter how much you do to stabilize the climate—even if you buy an electric vehicle, install solar panels, and reforest your backyard—your senses will not give you feedback on how you have helped.

Costa Rica faces the same challenge. People may notice the growing forests, but no one is going to see less carbon dioxide in the atmosphere. The reduced emissions are invisible and the effects are distant.

One way to make invisible and distant changes more noticeable is by imagining what they will look like. In experiments, people who interact with aged images of themselves save more money than those who do not. Extending this idea from saving money to saving earth, the practice of "visioning" rests on the logic that the more clearly we imagine the Anthropocene future we want, the more likely we will make it happen.

Here Costa Rica's goal to be carbon neutral by 2021 does double duty. It is essence that guides improvements to a complex system, and it is a statement of vision that renders those changes noticeable. Not only is Costa Rica reducing emissions, they are removing them, and they are going to keep subtracting until their country is carbon neutral. And not only are they going to be carbon neutral, they pledged to do so by 2021, long before any other nation.

The 2021 goal motivates Costa Ricans, and its noticeability inspires others and allows them to copy. If no one had copied Bruce Springsteen and Kate Orff, they would not have changed rock and roll and landscape architecture, and if no one copies Costa Rica, they won't change the Anthropocene. Even before they seriously began to pursue carbon neutrality, the nation was responsible for only 0.02 percent of global climate-changing emissions.

With their 2021 vision, though, Costa Rica has made their effort noticeable, and hopefully, copiable. Just as highway removals have spread since Sue Bierman's triumph, more ambitious climate visions are spreading. Germany, the United Kingdom, and the Netherlands are among the nations promising carbon neutrality by 2050. Amazon, the company, has committed to carbon neutrality by 2040 (Amazon, the rain forest, was already doing its part). With their bold vision, Costa Rica gave the world noticeable and copiable less.

A downside of bold visions is that the bolder they are, the harder they are to realize. And indeed, Costa Rica has moved back its 2021 goal to be among those promising neutrality by 2050. Still, the nation's bold imagination has brought real progress on the path to less: nearly all of its electricity comes from renewable sources; its forests continue to remove carbon; and—in a stop-doing that must become a trend—it has halted oil extraction. All this persisting to noticeable less has earned

Costa Rica recognition as the greenest country in the world, which has attracted a new stream of visitors, which is one of the ways the nation is making less pay.

The last step on the lesslist is to reuse our subtractions. As a doughnut hole reminds us, it's worth asking if we can reinvest what we take away.

Carbon dioxide pulled out of Costa Rica's atmosphere is turned into forests, thanks to good old photosynthesis. These forests, in turn, draw visitors from all over the world. And when my parents mail a picture of a rain forest sloth to Ezra, what I see is the improvement of another Anthropocene situation closely linked to climate change: the fact that half of all species face extinction in Ezra's lifetime.

Dr. Seuss hints at extinction in *The Lorax.* Ezra wonders, skeptically, how far the Humming-Fish will make it as they "walk on their fins and get woefully weary in search of some water that isn't so smeary." Our modern despoiling is worse than the Once-ler's. We humans have paved roads and cities, cleared forests, diverted rivers, and, for the coup de grâce, are warping the climate. Extinctions are a part of natural selection, yes, but their rates are now thousands of times higher than is normal. Mass extinctions destabilize entire ecosystems until, eventually, we are left without Truffulas—and everything else. But when carbon dioxide subtracted from the atmosphere is reused to make forests, there is more room for what (else) lives there.

The returns on Costa Rica's noticeable less come through handsomely in trees and other biodiversity. But what about the sector the Once-ler first fixated on—the economy?

Dr. Seuss's Thneeds are priced to rhyme at $3.98, and represent the massive amount of stuff we add to our world. His outlook on rampant consumerism is clear, as the Once-ler goes straight from cocky salesman—"A Thneed's a Fine-Something-That-All-People-Need"—to lonely has-been, with a shuttered factory in a place where the wind smells sour and only Grickle-grass grows. Even a preschooler knows this is not progress.

And yet, the whole Thneed folly would have increased gross domestic product (GDP). Lots of Thneeds were sold, a factory was built, and, for a time, "the whole Once-ler Family was working full tilt." GDP, our most influential, and blunt, measure of progress, loves more. Were the Truffula craze to happen in our own backyard, the market would judge it a success, Grickle-grass or no.

GDP hasn't always been such a powerful adding force. It was only during World War II, when there became a need to account for war spending, that what had been essentially a measure of workers' take-home pay was reconfigured as a measure of the amount of goods added. After the war, GDP's benchmark status spread right alongside more-ality. Countries requesting aid through the Marshall Plan had to produce an estimate of their GDP. The United Nations developed a template for member states to report theirs. Today, GDP growth is a goal for nearly every government.

GDP does what it was set up to do—but no more. It measures production, not welfare. As a result, it misses some useful less.

GDP counts the cost of the hospital visit but not the patient outcomes. A patient who has to stay in the hospital with a catheter infection runs up a higher bill than the patient who is treated and released. In a similar way, GDP struggles to measure advances in our ideas. No matter how much practical value we derive from internet searches, as long as our use of Google doesn't cost us anything, this act is excluded from GDP calculations. So is open-source work, like contributions to *Wikipedia*.

Not only does this metric miss useful less, it counts harmful more. The Once-ler's short-lived and disastrous exploitation of the Truffula valley grew GDP. So do hurricanes, oil spills, prison construction, and inefficient government spending.

Many of the things we want, from security, to healthy environments, to nurtured children, simply do not factor in this blunt national statistic—and may even lower it. Our Anthropocene accounting should reflect changes in quality of life and in the quality of our environment. That's easier written about than done.

Modifying measures requires that we ask difficult questions about the public good. Alternatives, including the Human Development Index put forth by the United Nations, are also not as simple to calculate as GDP. To measure human development, nations need to keep track of and then compile numerous indicators like life expectancy, education, and per capita income.

It is challenging to collect all of that information, but at least the indicators are agreed upon. Other aspects of human development, like psychological well-being and community life, are challenging to measure in the first place.

On the other hand, what we are talking about here is just a measurement. As a species smart enough to add reflective mirrors to outer space, we ought to be able to measure what matters.

Whereas GDP pushes us to add, other benchmarks can reward subtracting as well. Costa Rica assigns economic value to the carbon removal and storage provided by the nation's forests. By this standard, the Once-ler's Thneed factory would have to be indefinitely better than the Truffula valley, not just temporarily better than nothing. Using this modified measure, Costa Rica can put farmers to work growing trees, rather than cutting them down.

Perhaps Costa Ricans are just naive. This is a nation, after all, that got rid of its army in 1949, diverting war funds to education, health, pensions, and even an art museum. Its GDP is around $12,000 per person, whereas the United States' is around $65,000 per person. A less aggressive climate action plan might improve GDP, and a military surely would.

On the other hand, Costa Rica's reprioritized spending has improved literacy and health, to the point where its citizens now live longer lives, on average, than Americans. Maybe there is something to learn from their subtracting in the Anthropocene.

3.

Reforesting large swaths of Central America; setting a national vision for balanced adding *and* subtracting of carbon dioxide; placing a stop-doing on all oil extraction; changing measurements so that they notice progress through less—I'm well aware that these are not changes that any of us can make unilaterally. It is hard and strategic work that takes many hands. What's more, different situations demand different changes; instead of reforesting the midwestern United States, we might employ carbon-sequestering crops beneath wind turbines.

So what, exactly, should we leave as our Anthropocene legacy?

I appreciate the desire to be told, "Do this to fix climate change." If I thought there were one such thing to do, I most certainly wouldn't have waited until now to share. I would have put it in big bold font on the cover of this book and then again on every single page. That said, one thing I have learned is this: we need to leave options.

The Resilience Project was an unprecedented five-year partnership between top ecologists, economists, social scientists, and mathematicians from all over the world. Recognizing that our environmental crises are systemic, and respecting that systems are complex and unpredictable, the scientists did not conclude with misleadingly specific how-tos. Rather, they offered a general treatment: "We should strive to maximize the inherent potential of a system that is available for change, since that potential determines the range of future options possible."

Like Kate Orff, the scientists inverted, because "maximize inherent potential" sounds better than "subtract extraneous stuff." But to increase the amount of a city that is available for change means getting rid of the congesting freeway. To maximize the inherent potential of a society means subtracting racism. To leave the most useful collective knowledge requires pruning modern misconceptions. In other words, our Anthropocene remedy is to subtract.

Leaving options means cleaning up after ourselves. We can remove harmful distortions we have built into our systems. That is the good way to change things. Just as Leo Robinson, Run-DMC, and Ibram Kendi help us recognize and dismantle racism, we need to recognize and remove the invisible structures exploiting our common future.

We can divest, as Desmond Tutu is again recommending. The South African archbishop has already earned one Nobel Peace Prize for dismantling apartheid. He's setting himself up for a second as a vocal and influential advocate for using "the tactics that worked in South Africa against the worst carbon emitters." It's easier to dismantle apartheid after you have stopped investing billions of dollars propping it up; likewise, one good way to improve the planet is to stop funding its destruction. There is plenty of opportunity to do so. Right now, investments in fossil fuel companies represent more than five times as many emissions as scientists think it is safe to burn.

As with divestment from apartheid, the University of California system has been an early mover, becoming one of the largest institutions to divest their holdings in fossil fuel companies. And as with divestment from apartheid, the pope has called on all Catholics to do the same. A bishop among environmentalists, the author and activist Bill McKibben, has distilled the essence of his life's work down to getting more institutions and companies to divest. Ireland, the Gates Foundation, and hundreds of faith-based organizations are among those that have.

We leave options by subtracting harmful ideas and policies, and we leave options when we clean up our things. We want to be remembered and one way is by leaving behind physical things that we think will last beyond our lives. This is normal—but we have to be careful. If all we leave for future generations are second homes, monuments, and Exxon stock, then we are creating a Collyer brownstone at a planetary scale. To leave options, we need to subtract stuff.

Toward the end of Seuss's story, we learn that the elderly Once-ler has had a similar epiphany to the Resilience Project scientists about what to leave behind. To pass down his legacy to the boy who has come to see him, the Once-ler does not give him his factory or old machinery or any other artifact. To change the future ecological system, the Once-ler has realized that the best tool he can offer is in the last Truffula seed, which he gives to the boy. And to improve the future human system, the Once-ler leaves his tale. He still has an entrepreneur multitude; he charges the curious boy "fifteen cents, a nail, and the shell of a great-great-great-grandfather snail." But, for a fraction of the price of a Thneed, the boy gets wisdom that the Once-ler spent his whole life distilling.

As the Resilience Project scientists concluded, and as the Once-ler does, we should leave potential. By subtracting, we can leave a legacy of opportunities. Even growth-focused Truman acknowledged this point: "The material resources which we can afford to use for the assistance of other peoples are limited. But our imponderable resources in technical knowledge are constantly growing and are inexhaustible." Things are limited—but people needn't be.

8

From Information to Wisdom

Learning by Subtracting

1.

The Chinese philosopher Lao Tzu advised, "To attain knowledge add things every day. To attain wisdom subtract things every day."

Tzu's advice is thousands of years old, and timely. Just as subtracting is a survival skill for the Anthropocene, it is also a life hack for our information age.

Some downsides of being the latest generation to overlook Tzu's advice were outlined in Cal Newport's article "Is Email Making Professors Stupid?" which got to its intended audience via *The Chronicle of Higher Education*, and then Twitter, content-aggregating websites, and, yes, in professors' in-boxes. The gist of the article was that, while email is helpful for professors in some ways, the constant chatter has eroded precious time for uninterrupted thinking. Time to think is useful for many jobs—that's why Leslie Perlow tried to save her software engineers from time famine. But for professors, distraction-free time is the difference between doing our job and, well, stupid. To manage a deluge of messages, we sacrifice our ability to enlarge ideas.

The article hit my in-box three times—a downside of being affiliated with multiple departments. In each case, it provoked an immediate volley of emailed comments. Thankfully, before I chimed in on all three threads, an astute colleague pointed out the irony of professors discussing the article via email.

It's not just professors doing this to ourselves. The average American encounters one hundred thousand words a day, more than are in this book. In one internet minute (circa 2017), there were half a million tweets and more than three million Google searches. Our rate of email production? One hundred fifty-six million—per minute.

We all know the problem. We talk of information fatigue and overload. We may try information (a.k.a. "tech") diets. When our diets don't work, we may resort to an information purge. The phrase *too much information* is so common it has earned the acronym TMI, which has spread from texting into spoken language and even into old-media dictionaries. Ironically, TMI is itself information-dense—packing the same meaning in three keystrokes instead of twenty.

Too much information threatens our mental health, from the persistent frustration of interrupting emails, to the clinical anxiety born from an overload of shopping choices. Too much information endangers the participation required for a functioning democracy; people are inundated with so much content, good and bad, that it's hard to separate the signal from the noise. We can systematically consider the merits of every baby crib mattress, or learn the nuances of every candidates' plan—or frightening lack thereof—for responding to climate change. But we can't do it all.

There are real biological limits to how much information we can process at any given time. It's enough to have one face-to-face conversation at a time, but not two; somewhere between 60 and 120 bits per second (one character of text is loosely equivalent to 8 bits). Devoting our information processing to one situation means we can't use it elsewhere. Long before email, the satisficing authority Herbert Simon observed that "a wealth of information creates a poverty of attention."

Too much information clogging up mental bandwidth can also create other kinds of poverty. With their book, *Scarcity,* Eldar Shafir and Sendhil Mullainathan upended how we think about the relationship between economic poverty and bad decisions. Shafir, a psychology professor at Princeton, and Mullainathan, then an economics professor at Harvard, found that poor people are, indeed, more likely to make bad decisions. So far, so good for those who wish to blame poverty on the poor.

However, Shafir and Mullainathan show that this common inference should be reversed. It is not that bad decisions make one poor. It is that the cognitive effects of being poor lead to bad decisions. A high school student forced to spend her bandwidth thinking about whether to buy food for her younger siblings or books for her studies finds it harder to also be thinking about the content of those books. What's more, with her bandwidth already taxed by the books-or-food dilemma, she has less space to process new information about her situation, even if it is a program offering free books.

Poor people are often trapped in this condition of mental scarcity. The rest of us have the luxury of relieving it, as Lao Tzu and Herbert Simon recommend.

Information itself is no more the problem in our age than iron was in the Iron Age. It is a privilege and possibility. Perhaps the most hopeful news about our Anthropocene trajectory is that the growth of information outpaces the growth of energy use and climate changing emissions. Unlike fossil fuels or the climate commons, information is an inexhaustible resource. Still, for our information to improve our planet or anything else, we need to use it.

Modern adding tools, whether email or Twitter, don't help. But the feeling of too much information is not as new as we might think. Simon wrote that people were "completely maladapted to the world of broadcast systems and Xerox machines." Pre-Xerox, John Maynard Keynes, whose ideas helped usher in more-ality, pointed out that humankind would benefit from subtracting less useful information, which could then be compiled in what Keynes called an "accredited Index Expurgatorius," borrowing the name for the list of books that the Roman Catholic Church forbade its members to read unless certain passages were deleted or expurgated.

Way back when books had to be written and copied by hand, the Hebrew Bible was already warning that "there is no end to the making of many books, and much study wearies the body." Seneca, the Roman Stoic philosopher, deemed information so threatening that he devoted the second of his 124 end-of-life moral advice letters to "discursiveness in reading," warning that reading too many books can tax bandwidth one might otherwise use to confront new situations.

In her book *Too Much to Know,* the historian Ann Blair finds that we have always found ways to store, summarize, and sort information, and we are doing so now. In this sense, data servers aren't unlike Renaissance-era museums and libraries. *Wikipedia*-style summarizing traces its lineage to printed encyclopedias. Google, in sorting the world's information, is akin to alphabetizing, or the Dewey Decimal System.

Along with organizing information, we can also slow down the adding. Edward Tufte's teachings about post-satisficed information design eschew PowerPoint and its ease of adding slides, bullets, and other forms of non-information-ink. John McPhee, who has my vote and many others' for the best creative nonfiction writer ever, has never switched from a mechanical typewriter to word processing. Hemingway stuck with a pencil, a machine that slows the rate at which information is added, but still favors adding (the eraser is much shorter than the graphite).

Storing, summarizing, and sorting are not subtracting. Slowing is the information equivalent of Ben's no-bell. But Blair also unearthed one more historical strategy you will recognize: *select.* Like nature balancing adding and subtracting, we need to balance generating of information with selecting what is relevant and useful. Selecting relieves the tension between a wealth of information and poverty of attention.

Blair offers practical tips on this kind of triage. She recounts the selection filter used by the editors of *Encyclopédie,* a seventeen-volume alphabetical arrangement of Enlightenment ideas. The information contained, the authors decided, should be enough to rebuild society after a catastrophe. Information deemed unnecessary for that goal was subtracted. That is a high bar for relevance, but the principle is the same regardless of your goals.

Whether it is to rebuild society or to manage our in-box, selection requires that we distinguish ephemeral data from information. Most emails caught up in spam filters are clearly data, not information. In other cases, the distinction between data and information depends on the user. For most of my colleagues, an emailed warning to "whoever left their sandwich to grow mold in the break room refrigerator" is obviously just data. Others enjoy the sleuthing. The simplest selection filter is that, if you can't use it, it's definitely not information.

In other cases, the question is not data or information but whether the information is worth storing. Our subtracting skills can help make these decisions. Just as Ryan McFarland considered his two-year-old to subtract from the bike, considering the humans can guide us to data and even information we might subtract from our bandwidths, and from others'. We professors, for example, are quick to require students to learn new topics—but we often avoid hard questions about the relative value, beyond mental weight lifting, of existing ones. When we consider the students first, we force ourselves to ask these hard questions and to confront the truth that a poverty of attention is not good for learning. In extreme cases, too much information doesn't just tax students' bandwidth, it can make them believe cheating is their only option to meet the demands upon them. We've already seen how army officers will cut corners when they have too much to do, and students will do the same if they have too much to know.

Part of the reason it is so hard to subtract information, for professors and for others, is because we intuitively focus on the costs for the producer and the benefits for the users, respectively. This is the same two-sided thinking that Lego uses when figuring that they can make the 1,100-piece robot set for forty dollars (producer cost), and that parents will be willing to pay ninety dollars (value of user benefit) for it.

But in Legos and other transactions, the user also incurs costs. For material things, these costs are often negligible compared to what was paid for the good itself. I have to walk outside to recycle the Lego box, which is not my favorite activity, but not a significant cost.

When the transaction is for information, however, most of the cost to the user is not covered in the transaction itself. That's one reason I'm flattered that you have read this far. The cost of creating, collecting, writing, editing, marketing, and distributing the information in this book is one thing. You paid for that when you bought it. But the total cost of using the information also includes the time you've spent reading it. No matter how quickly you read, and whether you value your time at $15, $150, or $1,500 per hour, the value of the time you have invested in this book is worth many times what you paid for it.

To harness the benefits of our information privilege, we have to take ownership of the costs—both producer- and user-incurred. My professor friend who first emailed the "Is Email Making Professors Stupid?" article was understandably exasperated. Excess email is one of his sermons in faculty meetings and in hallway conversations. Yet look at what he did when he felt like his message wasn't getting across: he added information to the message. By forwarding the email, he created thirty-five new bandwidth-clogging in-box entries for the very group he was trying to protect.

The opportunity cost of information is not a new consideration either. Keynes's call for an Index Expurgatorius is nearly a century old, and since there have been libraries, a dark underbelly of librarianship is deciding which books to get rid of, to make room for new ones. Like Kate Orff cleaning, carving, and revealing, the librarians invert, not calling this change subtracting but instead "weeding."

If you're like me, weeding out a single book, even one you will never read, is traumatic. Librarians nonetheless know that weeding is necessary. The alternative is pulping: indiscriminately taking entire sections of books to a plant where they will be disassembled, dissolved into a milky liquid, and reconstituted as paper. When pulping took 210,000 nonfiction books from Manchester Central Library in the UK, the process brought librarians to tears. It was "cultural vandalism at industrial scale." Who's to say a book not checked out in the last ten years won't bring someone joy, or become the basis on which civilization is reconstructed? Weeding books is like our brain's synaptic pruning; pulping is a lobotomy.

Whether in our bookshelves, in-boxes, or brains, intentional and regular subtraction of information is far better than the alternative. If we don't sleep, which is when our synapses get pruned, our brains get overloaded and slow down. And if we don't consciously select information when we are awake, we end up with pulped classics, anxiety from information overload, and smart professors sending email about email making professors dumber.

The good news is that when we subtract information from our mental storerooms, our processing speeds up like a computer after closing a memory-intensive program that has been running in the background. Working at full capacity, we can create new knowledge—and perhaps even distill it into wisdom.

2.

I found courses in mechanics to be the highest hurdle en route to my undergraduate engineering degree. Mechanics is the branch of physics that deals with objects at rest and in motion, and it required me to go from plugging numbers into equations to visualizing how the concepts work in the world. It's a challenging leap, and without making it, I would have had no chance at succeeding in a bevy of later courses.

I hadn't made the leap heading into the third and final exam of my first mechanics course. I was carrying a C average. The teacher, Professor Viscomi, desperately wanted us to succeed, but we had to earn it. He had already taught thousands of students, twenty-five at a time, over three decades. It was well established that there would be no curving, no bonus points for attending office hours, and certainly no pleading for a grade adjustment.

For the first time in my life, I entered that third exam knowing that I was in danger of failing the course. If that happened, I would have two options: delay my degree progress (and ask my parents to pay extra tuition) to take the course again the following year, or change my major to something that didn't require me to pass mechanics.

Before handing back our exams, Professor Viscomi would ceremoniously write the highest and lowest grades on the chalkboard, leaving us all to guess who had earned them. After that make-or-break third exam, he wrote *98* and *47*.

Then he looked at me and smirked.

My classmates laughed at me and playfully jeered the student they assumed had gotten the 98. I was surprised, partly because I didn't expect Professor Viscomi would poke fun at my low grade, and partly because I didn't think I'd done that badly. Still, as he handed back our exams, I thought about majors that would give me more time for soccer, not to mention allow me to take courses that were not 80 percent male. That pleasant daydream ended before it got too far. When I received my test, I realized why Professor Viscomi had smirked. I had earned the 98.

What changed for me on that third exam? I had gone to class, studied, and done the practice problems for homework. But I had been doing those things all semester, and it had gotten me my C average. What changed was that, sometime before the third exam, I had mentally stripped the course down to its essence. I had found noticeable less.

For those with enough sense not to major in engineering, the first mechanics class (often called statics) boils down to applying Newton's second law of motion to objects that are either at rest or moving at constant velocity. All these scenarios can be described by one simple idea: force equals mass times acceleration, $F = ma$. I could derive everything I needed from that equation, the idea behind it, and a few rules for applying it to the analysis of loads acting on objects. Before that third exam, I stopped memorizing dozens of other equations and tangential ideas, which was the adding that had gotten me into the make-or-break predicament.

Just as I stopped worrying about derivable equations, you can stop thinking about everything you just read about mechanics and remember this: my breakthrough came by discarding less useful ideas. Professor Viscomi's course was daunting, but at its core, it was no different from any well-structured course. He taught many applications, but all were a derivation away from some basic ideas. I didn't need to know a bunch of forces and masses and accelerations, just $F = ma$.

No matter the subject, we build mental models as we learn. We take ideas and the relationships between them, and we use them to represent reality. To do so most effectively, we need to add *and* subtract, like nature, and like Maya Lin. Of course, we need to add detail to our mental models. But it is by stripping away elements of the system that we find essence. Weeding less useful ideas allows the indispensable ones to flourish.

It's best to subtract consciously, but we can also automate our information weeding, to avoid the need to pulp. One of my favorite less-is-more productivity tips is to write down fewer notes. This is the information version of the closet-cleaning tip in which you get rid of anything you haven't worn for a year. In both cases, you are forced to filter important content from inessential content. If the sweatshirt is worthy, I'll have worn it recently. If the idea deserves to make the cut, I'll have thought about it enough to remember without a notepad.

I still write some things down. I write down to-dos that keep me in the good graces of Monica at home and my colleagues at work, and I write notes to organize my thoughts. But I trust my brain for big ideas. I didn't need to write myself a reminder that "people overlook subtraction," because the finding kept coming up in my team's experiments. There was no way my microglial cells would prune that insight.

Since I've been writing fewer idea notes, I've saved frustrating time trying to make sense of old ideas my unconscious brain would have otherwise weeded. This has left more time for the good ideas. And I don't think I've forgotten anything important—though I suppose I wouldn't know.

3.

Pruning extraneous concepts helped me pass mechanics and then everything else. Filtering less useful information can protect our in-boxes and our bandwidth. That would be plenty, but let's persist with this less. Because if we can learn to subtract wrong ideas, we gain a rare power.

To appreciate why, let me tell you why Miss Carla never thanked me for dropping off three shiny new yellow dump trucks on Ezra's playground. Having recently turned two and a half, Ezra had just started at this "big-kids" preschool. His previous day care had an endless supply of dump trucks, but at Miss Carla's school, there was a single rusty truck, which had recently lost its dumper. I viewed this as a problem. As the bruises on their shins attest, two-and-a-half-year-olds, no matter the gender, love to race around pushing dump trucks. It's an activity perfectly suited to their unsteady walking skills and their instinct to show competence by moving big things.

Because preschool teachers are the highest form of humanity, Miss Carla not saying thank you spoke volumes. Clearly, I had done something very wrong. What I would eventually come to appreciate is that, by adding the dump trucks, I had thrown off her carefully crafted playground learning environment.

Miss Carla knew that two-and-a-half-year-olds loved dump trucks. That's why there was *one.* But she also knew that, if there were too many trucks, kids like Ezra would push them around all day long, as he had done at his previous school. If Ezra spent his days with a dump truck, he would miss out on the intentional playground litany of leftover wood decking, black drainage tubes, stackable milk crates, balls, mulch, sand, rocks, water, and, depending on the day, paper airplanes, bubbles, cloth wings, and baking-soda-and-vinegar volcanoes.

Education scholars have a name for how preschoolers, and the rest of us, learn. It's called *constructivism.* In short, we "construct" meaning based on the interactions between our surroundings and our minds. When he was behind a dump truck, Ezra only had one situation from which to build meaning. He didn't experience friction from slipping on the wet rocks. He didn't experience thermodynamics from how the sun's energy is absorbed inside the black drainage tubes, making them the warmest place to hide on the playground. He didn't experience gravity and off-center forces via falls from milk crate lookout towers. When Ezra's new playground led him to do all those things, he had many new physical experiences from which to build his knowledge.

The situations from which we build meaning need not be physical. A single dump truck requires sharing, or at least determining a pecking order—Ezra was second, and he made fast friends with Malcolm, who was first. Sharing (or not) was just the beginning. When puddles form on the beach-like sandbox that covers half of the playground, the preschoolers become little coordinating engineers, carving channels to move the water between the puddles.

Over time, I saw how the playground fostered behavior so advanced that it reminded me of the theory that monumental architecture preceded civilization, and not the other way around. On Potbelly Hill, the huge rock monuments are too big to have been carved and moved by a single band of hunter-gatherers. On Ezra's playground, the leftover wood decking pieces are too big to be maneuvered by a single preschooler. So, they work together.

No matter how many dump trucks we had on our playgrounds growing up, we eventually formed ideas about the laws of the universe. Knowledge comes courtesy of friends, teachers, and books about subtraction. Wherever the insights come from, the *construct* in constructivism refers to the fact that we build knowledge by adding new information to our preexisting ideas of how the world works. Our new ideas stand on old ones.

Individually, ideas shape our views of the world and of

ourselves. Collectively, knowledge construction is the cultural evolution that gives us our unique human advantages. As with building a civilization, we tend toward adding to build knowledge. Knowing something is better than not. But, as with building a civilization, the more ideas we have, the more chances and benefits there are to improve by removing.

The pinnacle of mental subtraction is when we remove ideas that are no longer correct, or that never were in the first place. This is not as simple as it sounds. Recognizing the value of removing wrong ideas so that new ones can be constructed on a more stable foundation, researchers have spent careers trying to figure out ways to do so.

It wasn't long ago that researchers described common but wrong interpretations of the world as "naive" physics, "naive" chemistry, or "naive" psychology—to be corrected, presumably, by encountering an "enlightened" scientist. Thousands of documented "misconceptions" distort how we learn about everything from gravity to climate change. If we could simply remove those wrong ideas like cancerous tumors, the new ones would fit in most naturally. It's a nice thought.

As we've learned, though, we neglect subtraction across the board, and our knowledge construction is no exception. Consider when Ezra got a prepackaged Lego fire truck from Santa Claus. This new experience conflicted with the knowledge he had spent much of his young life building. Ezra knew all about

Santa, his elves, and their wood workshop. And he knew that Santa did not have the plastic manufacturing capabilities necessary to make Legos.

So when Ezra asked, "How did Santa make my Legos?" I had to think fast.

"Oh, for stuff like Legos, Santa works directly with Amazon."

My son accepted my response in part because Santa outsourcing the Lego requests to an e-commerce company was knowledge that accommodated what he already knew about the world. But a more powerful factor was that my response didn't require him to subtract all the knowledge he had constructed about Santa Claus.

It's not just Santa's faithful who double down on their beliefs rather than let go. In 1954, the psychologist Leon Festinger, a student of Kurt Lewin, joined a cult that believed the world was going to end on December 21 of that year. The benefit of becoming a cult member was that, when visitors from outer space arrived at midnight before the apocalypse, believers would be escorted to a waiting spacecraft.

Festinger's clever plan had no downside. He would either have a fascinating psychological case study of how belief systems change when the world shows they are wrong. Or, he would be saved from the apocalypse.

The world didn't end. But, as with Ezra and Santa, the cult members ended up not being wrong. Festinger recalls how, in the minutes after midnight, the group debated which clock was the official timepiece of the apocalypse. Then they sat in silence as hours passed without any alien saviors. Just before five o'clock in the morning, the cult leader received great news. By sitting all night long, the group had convinced a higher power to save the world from destruction.

Festinger had his case study. The cult was an extreme example of the lengths we will go to avoid holding contradictory ideas in our own heads. Rather than resolve the conflict by subtracting the wrong idea, we bend both instead.

Chuckle all you want at four-year-olds and cult members, but we all resist shedding wrong ideas. Even the masterful Emerson, a man who spent his life thinking, fell prey. Recall how he characterized the ideas-to-things relationship: "The least enlargement of ideas . . . would cause the most striking changes of external things." For things, Emerson did not limit the kind of change. But for ideas, the only possibility was "enlargement."

Who knows whether Emerson actually failed to consider subtracting ideas, but the behavior is so prevalent that educators have given up trying to change it. "Misconceptions Reconceived," a pivotal paper published in 1994 in *The Journal of Learning Sciences,* put forth an alternative approach that has since become widely accepted. Its authors argued that we should focus on interrelationships among diverse pieces of knowledge, rather than try to find "particular flawed conceptions." In this approach, identifying the ideas students are building upon—whether right or wrong—remains essential. But rather than pruning misconceptions, the strategy is to consider them resources for cognitive growth.

There is much to like in this reconceiving of misconceptions. To focus on eradicating all ideas that diverge from the teacher's norm is to undervalue the diversity of ideas students bring to classrooms from their everyday experiences. This devaluing has fallen disproportionately on students from minority backgrounds, making them feel that their culture is not accepted or appreciated. With the reconceiving strategy, no longer is anyone "naive," which breaks down an artificial distinction between holders and learners of knowledge. Sure, instructors hold knowledge that learners do not. But the reverse is also true. There is no question that the reconceiving more accurately represents actual learning. The organization may change, but the elements must remain.

That said, the embrace of misconceptions does subtract one thing from the picture of how we learn: subtraction. Rather than replacement, which requires getting rid of one thing, *accommodation* has become the preferred term for describing the process of reframing our ideas to fit new experiences. Accommodation, not removal, is how we construct new knowledge.

Accommodating new ideas is better than rejecting them outright. Ezra did get closer to the truth when he realized that Santa's elves couldn't possibly mold plastic Legos by hand. When we don't remove any wrong ideas, though, we also run the risk of distorting the new ones.

4.

In his book *The Structure of Scientific Revolutions,* Thomas Kuhn distinguishes between ordinary and "revolutionary" progress in our collective ideas about the world. It's no surprise that academics like me find Kuhn's dissection of scientific epiphanies thrilling. As a dense treatise covering abstractions such as "exemplars," "incommensurability," and "paradigms," *The Structure of Scientific Revolutions* is an unlikely book to have earned its place among the most influential of the last century. But it has.

One revolution Kuhn studies is the transition from Aristotle's mechanics, in which all behavior was found in the object, to Galileo's view, which is closer to what I learned from Professor Viscomi because it accounts for forces like friction and gravity.

Such revolutions in ideas do more than change how we are taught science. They change how we see the world. Were Galileo and Aristotle to have the pleasure of watching Ezra pendulum back and forth on his playground swing, they would literally see different things. Galileo would see the inertia in Ezra and the forces acting on him: gravity, air resistance, me pushing, and Ezra pumping his legs. Aristotle would see a boy repeatedly drawn to his natural place at the center of the universe.

Through the lens of history, it becomes clear that profound advances in ideas require subtracting knowledge. Sure, Galileo had to add his new findings, but he couldn't simply build on Aristotle's. He had to identify, and dismantle, their very foundations.

Subtracting old ideas is vital to accepting the discoveries

we've encountered in this book. To use Elinor Ostrom's insights about the collective management of the commons requires getting rid of Hardin's tragedy. Seeing Herbert Simon's satisficing means discarding parts of the rational choice theory that preceded it. And to learn from Maya Lin's adding *and* subtracting requires that we stop thinking add or subtract.

Quantum theory revolutionary Max Planck held the pessimistic view that "a new scientific truth does not triumph by convincing its opponents and making them see the light, but rather because its opponents eventually die." It's true that additive accommodation is how learning happens, among preschoolers, cult members, and researchers. But it would be nice if we could get rid of wrong ideas without waiting for people, ourselves included, to die.

Fortunately, just because we don't tend to remove wrong ideas doesn't mean we can't. Whereas Kuhn focused on how revolutionary ideas find their way in the world, Nancy Nersessian examines how ideas form in a revolutionary thinker's head. To do this work, the Georgia Tech professor blends Kuhnian historical case studies with analysis using her expertise in cognitive science.

With this approach, Nersessian's work tells us more about the key subtracting battles in idea revolutions. There are the expected experiments and number crunching; but those are common skills. To generate novelty, to go from the known to the unknown, Nersessian finds that revolutionary scientists combine the highest levels of modern scientific practices with a humble and age-old device: analogies.

We all use analogies. They are convenient for describing concepts, as when I described Nersessian's case studies as "battles in idea revolutions."

Analogies can also teach new ideas. When we liken our brains' processing to a computer, or its pruning to an orchard, we build analogies from familiar things to better comprehend unknown ones. Even better, when we learn by analogy, research has found, what gets extended from one problem to another tends not to be the potentially distracting details but rather the essence. In other words, from the computer analogy, we do not assume our brains have a keypad and a sleek silver case with an Apple or Dell logo etched into them. But we do extend the computer's processing behavior as we comprehend our own brains. Analogies subtract detail to declutter the knowledge before it goes into our mental models.

In very special cases, as Nersessian shows, analogies can help us subtract wrong ideas. In these cases, analogies work because they feel like accommodation, in that they allow us to keep one foot in what we know while we seek new ground with the other. Studies of science learning show that presenting a new idea along with new evidence fails to remove misconceptions. The new idea plus the new evidence cannot overpower the embedded misconception. But if we take the same new idea and support it with an analogy to a valid idea that's already set in the learner's mind, then the misconception becomes vulnerable. A teacher can introduce Johannes Kepler's equations to describe planets' motion around the sun and describe all the places these equations hold true, but that may not be enough to remove a student's preconception that the earth is the center of the universe. If the student already understands how smaller electrons orbit around a larger atom, though, the teacher can extend that idea, via analogy, to the solar system, in which case the wrong idea about the earth-centric system, with the larger sun rotating around the smaller earth, might go away.

It's hard for a new idea to compete with preexisting ones, even when they are wrong. But a new idea buttressed by a correct preexisting idea can overcome a wrong preexisting idea. It even worked for Kepler himself, who, in describing his revolution, acknowledged: "I cherish more than anything else the Analogies, my most trustworthy masters."

Speaking of new ideas, over the last eight chapters, we have seen the untapped potential of subtracting and the rewards that can come with pursuing it. We have subtracted ideas, with the aid of analogy. We have removed the negative valence around taking away, because less is not a loss. No longer do we think add *or* subtract, because it should be add *and* subtract, like nature.

Persisting to noticeable less takes more mental steps, but I hope you're convinced that it's worth it. Subtracting can have outsize ripple effects. Sue Bierman's legacy lives on in her parks, in toddlers carrying monkey balloons, and whenever the Embarcadero story inspires another city dealing with a highway that takes more than it gives. Subtraction also really reverberates in our ideas; for this, let's turn to (thoughtfully) emailing professors.

"It's everywhere!" read the subject line of an email forwarded to me after I described subtraction neglect in a talk at Princeton University. The original email was sent from Robert Socolow, a physics professor who famously outlined climate stabilization wedges, to Elke Weber, who is among the most influential psychologists of the modern era and was my host for the day. What was in the body of that email between intellectual heavyweights? What great idea revolution had my talk inspired?

The email contained a recipe that had come across Socolow's online news feed. The recipe offered the secret to tasty roasted chicken: go heavy on the vinegar—and get rid of spices that dilute the effect.

We now know why that advice was counterintuitive, worthy of not only its place on Socolow's news feed but also the time he took to send the email. Because, as we've seen, we humans neglect an incredibly powerful option; we don't subtract. We pile on to-dos but don't consider stop-doings. We create incentives for high performance but don't get rid of obstacles to our goals. We draft new laws without abolishing outdated legislation. Whether we're seeking better behavior from our kids or designing new initiatives at work, we systematically opt for more over less.

Sure, I hope this book has convinced you that taking away can be incredibly powerful, rewarding, and fun. And I hope you are as inspired as I have been by *Subtract*'s heroes.

But my enduring wish is that, long after you forget about the subtracting exemplars, and even scientific details, you will see your world through a clearer lens. Whether it's roasted chicken, your to-do list, or a freeway blocking the best part of your city, I hope you will find more of the options you've been missing—and feel empowered to pursue them.

Takeaways

We have taken quite a journey together, time-traveling over millions of years, trespassing across scientific fields, and confronting overloaded situations from our in-boxes to the Anthropocene. Along the way, we discovered strong and reinforcing anti-subtraction forces. Yes, evolution can be a model for our conscious mental pruning; cultures of more make pocket parks possible; and our economic system gives divesting its power. Still, the deck is stacked against subtraction. Less is harder for us to imagine, whether in Legos, grids, or words. Even if we do manage to think of subtracting, innate adding leads to Collyer hoarding; instincts to show competence bring useless subfolders; cultures built by temples make freeways seem sacred; and modern more-ality favors addition to our homes and to our schedules.

Having seen all this adding, we began stripping it away. We found and shared noticeable less—we distilled a lesslist and other takeaways. Can a page of takeaways really sum up the lessons of an entire book? Isn't that the idea?

Here are your takeaways.

Invert: Try less before more. Subtract detail even before you act, as with triage. Then, once you are ready to make changes, put subtracting first—play Jenga. And remember, just because we now appreciate that less is not a loss, that does not mean that your audience and customers do. So, tell them about this book and, in the meantime, don't "subtract." Instead, clean, carve, and reveal. Add a unit of transformation.

Expand: Think add *and* subtract. Nature and Maya Lin show us that these are complementary approaches to change. Adding should cue subtracting, not rule it out. Try accessing a different multitude. The father might see what the bicycle designer misses. If you run out of multitudes, hire an editor. And don't forget to zoom out to see the field, because stop-doings and negative numbers are not unpossible. Plus, the field is where the tension is, and removing it is the "good" way to change systems. So sure, add diversity, but subtracting racism is the prize.

Distill: Focus in on the people. Bikes do not balance, but toddlers can. Strip down to what sparks joy. Decluttering delights, and so does the psychology of optimal experience. Use your innate sense for relative difference. Taking away a mammoth is a bigger transformation than adding one. Embrace complexity, but then strive for the essence. Forget objects, remember forces—and pass mechanics. Subtract information and accumulate wisdom.

Finally, persist: Keep subtracting. Can you make less undeniable? Bruce Springsteen made *Darkness* visible. Costa Rica made neutrality noticeable. Chip made an empty go-kart funny. Don't forget that you can reuse your subtractions, like doughnut holes. Subtract stuff to leave a legacy of options—like Sue, Leo, and Elinor.

I sincerely hope that you find yourself turning this book's ideas into better things, whatever they may be for you. Me? I can't wait to show Ezra's little sister the other way to play with Legos.

Acknowledgments

I am grateful for those who helped me see subtraction. For discovering the scientific essence of this book—Gabrielle Adams, Ben Converse, and Andy Hales. Our work together has been the highlight of my career. For finding the practical essence, while also elevating the science—Margo Fleming and the team at Brockman, Inc. The world of ideas is a better place for the work you do. For creating the scholarship and design, the science and art, that helped me hone these ideas—the hundreds of thinkers and doers in the book and its notes. Your work inspired me, and I hope to have paid it forward. And for supporting my trespassing between design and behavioral science—the University of Virginia, the National Science Foundation, and every student I've had the honor of working with.

I am just as thankful for those who helped me share subtraction. For making this book (and me) smarter and more concise—Meghan Houser and Sarah Murphy. I had high expectations for editors, and you both exceeded them. For removing barriers to sharing science—Amelia Possanza, Katherine Turro, Bob Miller, and the rest of the team at Flatiron Books. You set me free to learn, think, and write. For reading, discussing, and removing from early drafts—Morela Hernandez, Jennie Chiu, Christine Moskell, Lucca Huang, Ted Burns, Evan Nesterak, Dave Nussbaum, and Heather Kreidler. If only all my writing could be distilled through such an impressive group. And finally, for helping me and this book stay focused through postpartum and a pandemic—my parents, Laurie and Larry Klotz, and my partner, Monica Patterson.

Image Credits

Figure 1: Created by Andy Hales

Figure 2: Photo by Elliott Prpich

Figure 4: Hanne Huygelier, Ruth Van der Hallen, Johan Wagemans, Lee de-Wit, and Rebecca Chamberlain, "The Leuven Embedded Figures Test (L-EFT): Measuring Perception, Intelligence or Executive Function?" PeerJ 6:e4534. https://doi:10.7717/peerj.4524.

Figure 5: (Fgrammen, "File:Savannah-four-wards.png," Wikimedia Commons, https://commons.wikimedia.org/w/index.php?curid=19978483)

Figure 6: Created by Maya Lin

Figure 7: Photo by Nancy Perkins

Figure 11: Town Branch Commons, Lexington, Kentucky. Image of Winning Competition Entry, 2013. SCAPE / LANDSCAPE ARCHITECTURE

Notes

Introduction: The Other Kind of Change

1 *federal support for highways:* Richard F. Weingroff, "Federal-Aid Highway Act of 1956: Creating The Interstate System," *Public Roads* 60, no. 1 (1996).

2 *the Embarcadero Freeway:* Charles Siegel, *Removing Freeways—Restoring Cities* (n.p.: Preservation Institute, 2007), ebook.

3 *Loma Prieta earthquake:* Hai S. Lew, "Performance of Structures During the Loma Prieta Earthquake of October 17, 1989," NIST Special Publication, 778 (Gaithersburg, MD: National Institute of Standards and Technology, 1990).

4 *most expensive earthquake:* Gregory Wallace, "The 10 Most Expensive U.S. Earthquakes," CNN Business, August 25, 2014.

4 *post-earthquake freeway had been rendered unusable:* Stephan Hastrup, "Battle for a Neighborhood," *Places* 18, no. 2 (2006): 66–71.

4 *far more than knocking it down:* "Embarcadero Freeway," Congress for the New Urbanism (CNU), https://www.cnu.org/what-we-do/build-great-places/embarcadero-freeway.

4 *Cypress Street Viaduct collapsed:* Douglas Nims et al., "Collapse of the Cypress Street Viaduct as a Result of the Loma Prieta Earthquake," Earthquake Engineering Research Center, University of California, UCB/EERC 89/16 (Berkeley, CA, 1989).

4 *Local businesses agreed:* Mark A. Stein and Norma Kaufman, "Future of Embarcadero Freeway Divides San Francisco," *Los Angeles Times,* April 13, 1990.

4 *Herb Caen:* Siegel, *Removing Freeways.*

5 *city's board of supervisors:* "Resolution Endorsing the Concept of a Subsurface Freeway on the Embarcadero Subject to Conditions," *Journal of Proceedings, Board of Supervisors, City and County of San Francisco* 52, no. 1 (April 16, 1990), 405.

5 *The decade after removal saw:* Robert Cervero et al., "From Elevated Freeways to Surface Boulevards: Neighborhood and Housing Price Impacts in San Francisco," *Journal of Urbanism* 2, no. 1 (March 2009): 31–50.

5 *Trips were rerouted:* Ibid.

5 *as many walkers as it does riders:* Bryan Goebel, "Bikeway on San Francisco's Embarcadero a Step Closer to Reality," KQED, July 24, 2014.

5 San Francisco Chronicle *was reporting:* Edward Epstein, "Ceremony Opens an Era of Optimism for S.F. Embarcadero," *SFGate,* June 17, 2000.

6 *"the quintessential neighborhood activist":* Randy Shaw, "Sue Bierman: Neighborhood Activist Led Battles Against San Francisco's Runaway Development," *BeyondChron,* August 9, 2006.

6 *Leo Robinson:* International Longshore and Warehouse Union (ILWU), "Leo Robinson: ILWU Activist Led Anti-Apartheid Struggle," *Dispatcher,* January 30, 2013.

7 *steel, auto parts, and wine:* Peter Cole, "No Justice, No Ships Get Loaded," *International Review of Social History* 58, no. 2 (August 2013): 185–217.

7 *city of Oakland had pulled:* Bay Area Free South Africa Movement, "Oakland: Divest Now!," 1985 flyer, http://www.freedomarchives.org/Documents/Finder/DOC54_scans/54.OaklandDivestNow.flyer.pdf.

7 *state of California followed Oakland's lead:* Robert Lindsey, "California's Tough Line on Apartheid," *New York Times,* August 31, 1986.

8 *when Mandela spoke in Oakland:* Nelson Mandela, June 30, 1990, Oakland-Alameda County Coliseum, Oakland, CA.

8 *Noble Prize winners such as herself:* "Elinor Ostrom—Facts," Nobel Media AB, https://www.nobelprize.org/prizes/economic-sciences/2009/ostrom/lecture/.

8 *influential 1968 essay:* Garrett Hardin, "The Tragedy of the Commons," *Science* 162, no. 3859 (1968): 1243–48.

9 *Hardin argued:* Garrett Hardin, "Lifeboat Ethics: The Case Against Helping the Poor," *Psychology Today* 8 (1974): 38–43. In their eighties, Hardin and his ALS-stricken wife committed suicide. They were survived by four children, presumably safely in the lifeboat.

9 *Ostrom showed that these assumptions:* Elinor Ostrom, *Governing the Commons: The Evolution of Institutions for Collective Action* (Cambridge, UK: Cambridge University Press, 1990).

9 *Indonesian forests:* Elinor Ostrom, "Self-Governance and Forest Resource," Center for International Forestry Research, Occasional Paper no. 2 (February 1999): 1–19.

9 *Nepalese irrigation systems:* Elinor Ostrom, *Crafting Institutions for Self-Governing Irrigation Systems* (San Francisco: Institute for Contemporary Studies and Center for Self-Governance, 1992).

9 *New England lobster fisheries:* Paul Dragos Aligica and Ion Sterpan, "Governing the Fisheries: Insights from Elinor Ostrom's Work," in *Institutions and Policies*, ed. R. Wellings (London: Institute of Economic Affairs Monographs, 2017).

10 *more like a drama:* National Research Council, *The Drama of the Commons* (Washington, D.C.: National Academies Press, 2002).

12 *a quarter of a million items:* Mary MacVean, "For Many People, Gathering Possessions Is Just the Stuff of Life," *Los Angeles Times,* March 21, 2014.

13 *federal regulations that are twenty times as long:* "Reg Stats: Total Pages in the Code of Federal Regulations and the Federal Register," GW Regulatory Studies Center, https://regulatorystudies.columbian.gwu.edu/reg-stats.

13 *describe learning as "knowledge* construction": National Research Council, *How People Learn: Brain, Mind, Experience, and School* (Washington, D.C.: National Academies Press, 2000).

16 *elegance:* Matthew E. May, *In Pursuit of Elegance: Why the Best Ideas Have Something Missing* (New York: Broadway Books, 2010).

16 *"less is more":* Rory Stott, "Spotlight: Mies van der Rohe," *ArchDaily,* March 27, 2020. An architect, Ludwig Mies van der Rohe (1886–1969), adopted the motto "Less is more" to describe his aesthetic turned platitude.

17 *Cal Newport preaches digital minimalism:* Cal Newport, *Digital Minimalism: Choosing a Focused Life in a Noisy World* (New York: Portfolio /Penguin, 2019).

17 *chef Jamie Oliver distills recipes:* Jamie Oliver, *5 Ingredients: Quick & Easy Food* (New York: Flatiron Books, 2019).

17 *Marie Kondo declutters homes:* Marie Kondo, *The Life-Changing Magic of Tidying Up* (Berkeley, CA: Ten Speed Press, 2014).

18 *Ralph Waldo Emerson:* Ralph Waldo Emerson, *The Complete Works of Ralph Waldo Emerson: Miscellanies,* vol. XI (Cambridge, MA: Riverside Press, 1904), 164–66.

18 *William James:* William James, *The Principles of Psychology,* (New York: Henry Holt and Company, 1890).

19 *carbon dioxide emissions are veering downward:* Renee Cho, "COVID-19's Long-Term Effects on Climate Change—For Better or Worse," State of the Planet, Earth Institute, Columbia University, online, June 25, 2020.

19 *African Americans are three times:* Centers for Disease Control and Prevention, "COVID-19 Hospitalization and Death by Race/Ethnicity," Cases, Data & Surveillance, August 18, 2020.

19 *redlining legacy that confines:* Rashawn Ray, "Why Are Blacks Dying at Higher Rates from COVID-19?," Brookings Institution, April 9, 2020.

Chapter 1: Overlooking Less: Legos, the Lab, and Beyond

23 *produce fewer climate-changing emissions:* Buildings alone account for more emissions than automobiles and airplanes combined. See Our World in Data, "Greenhouse Gas Emissions by Sector, World, 2016," https://ourworldindata.org/grapher/ghg-emissions-by-sector.

24 *anchoring on irrelevant numbers:* Leidy Klotz et al., "Unintended Anchors: Building Rating Systems and Energy Performance Goals for U.S. Buildings," *Energy Policy* 38, no. 7 (July 2010): 3557–66.

24 *unthinkingly accepting default choices:* Tripp Shealy et al., "Using Framing Effects to Inform More Sustainable Infrastructure Design Decisions," *Journal of Construction Engineering and Management* 142, no. 9 (September 2016).

24 *swayed by examples:* Nora Harris et al., "How Exposure to "'Role Model' Projects Can Lead to Decisions for More Sustainable Infrastructure," *Sustainability* 8, no. 130 (2016).

25 *"Tim Ferriss's* 4-Hour Workweek": Timothy Ferriss, *The 4-Hour Workweek* (London: Vermillion, 2010).

25 *"best professor under 40":* Maya Itah, "Best 40-Under-40 Professor Gabrielle Adams," *Poets&Quants,* February 12, 2014.

26 *"marshmallow test" of delayed gratification:* Walter Mischel and Ebbe E. Ebbesen, "Attention in Delay of Gratification," *Journal of Personality and Social Psychology* 16, no. 2 (1970): 329–37.

26 *scored higher on the SATs:* Walter Mischel, *The Marshmallow Test: Mastering Self-Control* (Boston: Little, Brown, 2014).

27 *We had research assistants recruit passersby:* Once Gabe, Ben, Andy, and I dreamed up a scenario, our research assistants would help make it happen. They'd set up their tables, sometimes by the food trucks on campus, and sometimes among the restaurants and bars adjacent to it. Then our assistants would watch people from all walks of life play with Legos and more, all the while carefully recording exactly what happened over thousands of iterations.

33 *judgment and decision-making conference:* Annual Meeting of the Society for Experimental Social Psychology, Seattle, WA, October 4–6, 2018.

33 *one of my role models:* Elke Weber, Princeton psychology professor after this talk: Leidy Klotz, "Saving Carbon Where Design Meets Psychology," David Bradford Energy and Environmental Policy Seminar Series, Princeton University, October 1, 2018.

34 *the IKEA effect:* Michael I. Norton et al., "The IKEA Effect: When Labor Leads to Love," *Journal of Consumer Psychology* 22, no. 3 (July 2012): 453–60.

34 *previous additions are sunk costs:* Hal R. Arkes and Catherine Blumer, "The Psychology of Sunk Cost," *Organizational Behavior and Human Decision Processes* 25, no. 1 (February 1985): 124–40.

34 *something exists, there's a good reason:* Scott Eidelman et al., "The Existence Bias," *Journal of Personality and Social Psychology* 97, no. 5 (November 2009): 765–75.

34 *losses loom larger than gains:* Daniel Kahneman and Amos Tversky, "An Analysis of Decision Under Risk," *Econometrica* 47, no. 2 (March 1979): 263–91.

35 *"accessibility":* A couple of good reviews of what we are calling accessibility: Shelley E. Taylor et al., "Salience, Attention, and Attribution: Top of the Head Phenomena," *Advances in Experimental Social Psychology* 11 (1978): 249–88; see also E. Tory Higgins, "Knowledge Activation: Accessibility, Applicability, and Salience" in *Social Psychology: Handbook of Basic Principles,* ed. E. Tory Higgins and Arie W. Kruglanski (New York: Guilford Press, 1996).

37 *Accessibility can also lead us astray:* Allen Newell and Herbert A. Simon, *Human Problem Solving* (Englewood Cliffs, NJ: Prentice-Hall, 1972).

37 *an architecture professor:* Matthew Jull, University of Virginia, after this talk: Leidy Klotz, "Design and Human Behavior," A Convergence Dialogue, UVA Environmental Resilience Institute, September 21, 2018.

42 *mental processing power faces competing demands:* Anuh K. Shah et al., "Some Consequences of Having Too Little," *Science* 338 (November 2012): 682–85.

42 *leaves less brainpower:* Anandi Mani et al., "Poverty Impedes Cognitive Function," *Science* 341, no. 6149 (August 2013): 976–80.

46 *we're missing ways to make:* Gabe Adams et al., "People Systematically Overlook Subtractive Changes," *Nature* 592 (April 2021): 258–61.

Chapter 2: The Biology of More: Our Adding Instincts

47 *Bowerbirds:* Gerald Borgia, "Sexual Selection in Bowerbirds," *Scientific American* 254, no. 6 (June 1986): 92–100.

49 *key idea with one word,* competence: Robert W. White, "Motivation Reconsidered: The Concept of Competence," *Psychological Review* 66, no. 5 (1959): 297–333.

49 *Albert Bandura extended White's idea:* Albert Bandura, "Self-Efficacy: Toward a Unifying Theory of Behavioral Change," *Psychological Review* 84, no. 2 (March 1977): 191–215.

50 *evolution created my research team:* Or, as evolutionary biologist and author Richard Dawkins put it: "A plane or a car is explained by a designer himself, the engineer, but that's because the designer is explained by natural selection." See Patrick Richmond, "Richard Dawkins' Darwinian Objection to Unexplained Complexity in God," *Science & Christian Belief* 19, no. 2 (2007): 101.

51 *mating between humans and Neanderthals:* Scott H., "Finding Your Inner Neanderthal," *23andMe Blog,* December 11, 2011. For details on the current version, including cites to major peer-reviewed articles, see also Robert P. Smith et al., "Neanderthal Ancestry Inference," 23andMe, white paper 23-05 (December 2015).

51 *a crucial time in human history:* There is some debate about how abrupt/revolutionary these changes were. For the faster view, see Richard G. Klein, "Anatomy, Behavior, and Modern Human Origins," *Journal of World Prehistory* 9, no. 2 (1995): 167–98. The more gradual view is explained in Sally McBrearty and Allison S. Brooks, "The Revolution That Wasn't: A New Interpretation of the Origin of Modern Human Behavior," *Journal of Human Evolution* 39, no. 5 (November 2000): 453–563. The point for us is that humans have not always had these skills.

51 *our ancestors developed new abilities:* Thomas Wynn and Frederick L. Coolidge, "The Implications of the Working Memory for the Evolution of Modern Cognition," *International Journal of Evolutionary Biology* 741357 (2011): 1–12.

51 *were all hunter-gatherers:* Richard B. Lee and Richard Daly, *The Cambridge Encyclopedia of Hunters and Gatherers* (Cambridge, UK: Cambridge University Press, 1999).

52 *health can even depend on eating:* World Health Organization, "Obesity and Overweight," WHO Fact Sheets, April 1, 2020.

53 *The Collyer brothers:* William Bryk, "The Collyer Brothers," *New York Sun,* April 13, 2005. See also "The Collyer Mystery Solved—Langley Kept Faith with Brother to the End, Died Under Junk Near Him in Their 'Castle,'" *Pittsburgh Press,* April 9, 1947, 21. See also Franz Lidz, *Ghostly Men: The Strange but True Story of the Collyer Brothers and My Uncle Arthur, New York's Greatest Hoarders* (New York: Bloomsbury, 2003).

53 *decades worth of adding:* "Weird Tales," *Reading Eagle,* August 7, 1942, 6.

54 *"acquisitiveness," or how and why:* Stephanie D. Preston et al., *The Interdisciplinary Science of Consumption* (Cambridge, MA: MIT Press, 2014). This underappreciated book is a must-read for anyone studying issues, from poverty to water scarcity, with roots in human behavior and resource constraints.

54 *an object acquisition task:* Stephanie D. Preston et al., "Investigating the Mechanisms of Hoarding from an Experimental Perspective," *Depression and Anxiety* 26 (2009): 425–37.

55 *stress correlates with adding objects:* Brian D. Vickers and Stephanie D. Preston, "The Economics of Hoarding," in *Oxford Library of Psychology: The Oxford Handbook of Hoarding and Acquiring* (Oxford, UK: Oxford University Press, 2014), 221–32.

56 *demise of the Collyer brothers:* It wasn't until 2013 that hoarding was officially classified as a disorder. So, Langley was never officially diagnosed or treated. See American Psychiatric Association, *Diagnostic and Statistical Manual of Mental Disorders, Fifth Edition (DSM-5)* (Washington, D.C.: American Psychiatric Press, 2013).

56 *kangaroo rats had their piles:* Stephanie D. Preston and Lucia F. Jacobs, "Conspecific Pilferage but Not Presence Affects Merriam's Kangaroo Rat Cache Strategy," *Behavioral Ecology* 12, no. 5 (September 2001): 517–23.

56 *stockpiling has been observed:* Erin Keen-Rhinehart et al., "Psychological Mechanisms for Food-Hoarding Motivation in Animals," *Philosophical Transactions of the Royal Society of London. Series B, Biological Sciences* 365, no. 1542 (March 2010): 961–75.

57 *behavior is probably a biological instinct:* Nasir Naqvi et al., "The Role of Emotion in Decision Making: A Cognitive Neuroscience Perspective," *Current Directions in Psychological Science* 15, no. 5 (October 2006). See also John P. O'Doherty, "Reward Representations and Reward-Related Learning in the Human Brain: Insights from Neuroimaging," *Current Opinion in Neurobiology* 14, no. 6 (December 2004): 769–76. See also Wolfram Schultz et al., "Reward Processing in Primate Orbitofrontal Cortex and Basal Ganglia," *Cerebral Cortex* 10, no. 3 (2000): 272–83.

57 *neuroscientists have confirmed:* Ming Hsu et al., "Neural Systems Responding to Degrees of Uncertainty in Human Decision-Making," *Science* 310, no. 5754 (December 2005): 1680–83. See also Brian Knutson et al., "Neural Predictors of Purchases," *Neuron* 53, no. 1 (January 2007): 147–56. See also Steven W. Anderson et al., "A Neural Basis for Collecting Behaviour in Humans," *Brain* 128 (2005): 201–12. See also Ian Q. Whishaw and Rick A. Kornelsen, "Two Types of Motivation Revealed by Ibotenic Acid Nucleus Accumbens Lesions: Dissociation of Food Carrying and Hoarding and the Role of Primary and Incentive Motivation," *Behavioural Brain Research* 55, no. 2 (1993): 283–95.

57 *keep us clicking and scrolling:* Daria J. Kuss and Mark D. Griffiths, "Internet and Gaming Addiction: A Systematic Literature Review of Neuroimaging Studies," *Brain Science* 2, no. 3 (September 2012): 347–74.

58 *treatment options, like cognitive behavioral therapy:* "Hoarding Disorder," Mayo Clinic, February 3, 2018, https://www.mayoclinic.org/diseases-conditions/hoarding-disorder/diagnosis-treatment/drc-20356062.

58 *genetics, as opposed to education:* Elizabeth S. Spelke, "Nature, Nurture, and Development," in *Handbook of Perception and Cognition: Perception and Cognition at Century's End,* 2nd ed. (San Diego, CA: Academic Press, 1998), 333–371.

59 *research most relevant to Ezra's peas:* Camilla K. Gilmore et al., "Symbolic Arithmetic Knowledge Without Instruction," *Nature* 447, no. 31 (May 2007): 589–91.

59 *didn't know how to do arithmetic:* Random chance would predict that they would be right only half of the time. But the kids did much better than that, correctly picking who had more candies nearly three-quarters of the time.

60 *sense for less and more:* Gilmore et al., "Symbolic Arithmetic."

60 *pervasiveness of this innate sense:* Stanislas Dehaene, *The Number Sense: How the Mind Creates Mathematics* (Oxford, UK: Oxford University Press, 1997).

61 *"just a cog wheel":* Ibid., 28.

61 *specific brain networks activate:* The angular gyrus and the horizontal part of the intraparietal sulcus, see Wim Fias et al., "Processing of Abstract Ordinal Knowledge in the Horizontal Segment of the Intraparietal Sulcus," *Journal of Neuroscience* 27, no. 33 (August 2007): 8952–56. See also Mohamed L. Seghier, "The Angular Gyrus: Multiple Functions and Multiple Subdivisions," *Neuroscientist* 19, no. 1 (February 2013): 43–61.

61 *engrave notches in baboon bones:* One of the earliest known mathematical artifacts, dating from around forty thousand years ago, is the Lebombo bone, a baboon fibula with twenty-nine notches. See Johanna Pejlare and Kajsa Brating, "Writing the History of Mathematics: Interpretations of the Mathematics of the Past and Its Relation to the Mathematics of Today," in *Handbook of the Mathematics of the Arts and Sciences* (Switzerland: Springer Nature, 2019).

62 *change we sense depends on:* Editors of the Encyclopedia Britannica, "Weber's Law," *Encyclopedia Britannica,* January 31, 2020.

64 *Pascal overlooked one type of less:* Blaise Pascal, *Pascal's Pensées* (New York: E. P. Dutton, 1958).

64 *negative numbers had never been:* Lisa Hefendehl-Hebeker, "Negative Numbers: Obstacles in Their Evolution from Intuitive to Intellectual Constructs," *For the Learning of Mathematics* 11, no. 1 (February 1991): 26–32.

66 *picture a spatial number line:* Edward M. Hubbard et al., "Interactions Between Number and Space in Parietal Cortex," *Nature Reviews Neuroscience* 6 (June 2005): 435–48.

69 *term for lithic reduction:* William Andrefsky Jr., *Lithics: Macroscopic Approaches to Analysis,* 2nd ed., Cambridge Manuals in Archaeology (Cambridge, UK: Cambridge University Press, 2005).

70 *what the rock would look like:* K. Kris Hirst, "Levallois Technique—Middle Paleolithic Stone Tool Working: Advanced in Human Stone Tool Technology," ThoughtCo, May 30, 2019.

70 *more likely to pass down our genes:* Understanding Evolution Team, "Welcome to Evolution 101!," Understanding Evolution, https://evolution.berkeley.edu/evolibrary/article/evo_01.

71 *modern brains are smaller than Neanderthals':* Keely Clinton, "Average Cranium / Brain Size of Homo neanderthalensis vs. Homo sapiens," Montague Cobb Research Lab, Howard University, December 24, 2015. See also, Bridget Alex, "Neanderthal Brains: Bigger, Not Necessarily Better," *Discover Magazine,* September 21, 2018.

71 *our brain cells shrink:* Christopher Wanjek, "Sleep Shrinks the Brain—and That's a Good Thing," *LiveScience,* February 3, 2017.

71 *microglial cells:* Katrin Kierdorf and Marco Prinz, "Microglia in Steady State," *Journal of Clinical Investigation* 127, no. 9 (September 2017): 3201–9.

Chapter 3: The Temple and the City: Adding Brings Civilization, and Civilization Brings More

74 *no longer the sole occupation:* Joshua J. Mark, "Daily Life in Ancient Mesopotamia," *Ancient History Encyclopedia,* April 15, 2014.

75 *Antoine de Saint-Exupéry, observed:* Antoine de Saint-Exupéry, trans. Lewis Galantiere, *Wind, Sand and Stars* (New York: Houghton Mifflin Harcourt, 1967).

75 *cultures and civilizations are like:* Peter J. Richerson and Robert Boyd, *Not by Genes Alone: How Culture Transformed Human Evolution* (Chicago: University of Chicago Press, 2005).

75 *it does so much faster:* Of course, cultural reasons for behavior are intertwined with biological ones. One reason cities and the resulting cultures came to be in the first place is because working together helped people pass on their genes. The relationship works in the other direction too. Some cultural adaptations can change the human gene pool, as when horse-riding invaders from the Russian steppe overwhelmed horseless defenders of the first cities.

76 *record of what we did:* In his pioneering work, *The City in History: Its Origins, Its Transformations, and Its Prospects* (1961), 119, sociologist Lewis Mumford distinguishes cities by their "ability to transmit in symbolic

forms and human patterns a representative portion of a culture." Providing a record of collective behavior, including transformations, is not just something cities happen to do; it is what makes them cities.

77 *Coba's pyramid:* George E. Stuart, "Coba," in *The Oxford Encyclopedia of Mesoamerican Cultures,* ed. Davíd Carrasco (Oxford, UK: Oxford University Press, 2001).

77 *what counts as monumental architecture:* Bruce G. Trigger, "Monumental Architecture: A Thermodynamic Explanation of Symbolic Behavior," *World Archaeology* 22, no. 2 (October 1990): 119.

78 *home to about thirty thousand people:* U.S. Census Bureau, *1830 Census: Abstract of the Returns of the Fifth Census* (Washington, D.C.: Duff Green, 1832).

79 *perfectly good one about thirty-five miles:* Smithsonian Institution Research Information System, "Washington Monument," *Inventory of American Sculpture,* IAS 75006044.

79 *design competition went ahead:* It was no surprise that Robert Mills won the competition for the monument: he had already designed the Washington Monument in neighboring Baltimore.

79 *nearly quadrupled its population:* From 16 million in 1836 to 61.4 million in 1888 according to Max Roser, Hannah Ritchie, and Esteban Ortiz-Ospina, "World Population Growth," Our World in Data, https://ourworldindata.org/world-population-growth.

80 *Potbelly Hill:* Andrew Curry, "Gobekli Tepe: The World's First Temple?" *Smithsonian Magazine,* November 2008.

80 *as tall as giraffes, and more:* The tallest pillars are sixteen feet weighing 7–10 tons, while giraffes are fourteen to nineteen feet and over 1,750 pounds according to "Giraffe," *National Geographic,* https://www.nationalgeographic.com/animals/mammals/g/giraffe/.

81 *new-and-improved theory:* "Zuerst Kam Der Tempel, Dann Die Stadt," *Istanbuler Mitteilungen* 50 (2000): 5–41.

81 *is increasingly believed:* For example, Ian Hodder, a Stanford University professor of anthropology who has directed digs elsewhere points out that "Gobekli changes everything. It's elaborate, it's complex and it is pre-agricultural." See Nicholas Birch, "Oldest Temple on Earth Discovered in Turkey," *Eurasianet,* May 5, 2008.

83 *theory now gaining momentum:* For an earlier interdisciplinary view, see Robin Dunbar et al., *The Evolution of Culture* (New Brunswick, NJ: Rutgers University Press, 1999).

83 *thus growing civilization:* Robin Dunbar et al., *Social Brain, Disturbed Mind* (Oxford, UK: Proceedings of the British Academy, 2010).

85 *often firsthand account:* Markus has written hundreds of papers, which have been cited more than one hundred thousand times.

85 *research on cultural worldviews:* Hazel Rose Markus and Alana Conner, *Clash!: How to Thrive in a Multicultural World* (New York: Plume, 2014).

85 *perpetuate these "interdependent" or "independent" views:* Hazel R. Markus and Shinobu Kitayama, "Culture and the Self: Implications for Cognition, Emotion, and Motivation," *Psychological Review* 98, no. 2 (April 1991): 224–53.

87 *by changing his subjects' perspectives—literally:* Bill Reiche, "How Do You Know Which Way is Up?," *Popular Science,* December 1950, 109–13.

87 *"field dependence":* Developed by Herman A. Witkin et al., *Psychological Differentiation* (New York: John Wiley & Sons, 1962).

88 *evidence of systematic variation:* Solomon E. Asch and Herman A. Witkin, "Studies in Space Orientation I–IV," *Journal of Experimental Psychology,* 1948.

88 *Embedded Figures Test:* Herman A. Witkin, "Individual Differences in Ease of Perception of Embedded Figures," *Journal of Personality* 19, no. 1 (1950).

89 *versions of this ingenious test:* Mutsumi Imai and Dedre Gentner, "A Cross-Linguistic Study of Early Word Meaning: Universal Ontology and Linguistic Influence," *Cognition* 62 (1997): 169–200.

91 *James Oglethorpe:* Thomas D. Wilson, *The Oglethorpe Plan: Enlightenment Design in Savannah and Beyond* (Charlottesville: University of Virginia Press, 2012).

93 *turn vacant lots into "pocket" parks:* American Society of Planning Officials (ASPO), "Vest Pocket Parks," Information Report No. 229 (Chicago, December 1967).

93 *try out this idea in Harlem:* "Collyer Brothers Park," NYC Parks, https://www.nycgovparks.org/about/history/historical-signs/listings?id=7845.

94 *pocket parks that they require:* ASPO, "Vest Pocket Parks."

96 *original description of her design:* "Designer of the Vietnam Veterans Memorial: Maya Lin," Vietnam Veterans Memorial Fund, https://www.vvmf.org/About-The-Wall/history-of-the-vietnam-veterans-memorial/Maya-Lin/.

96 *how she approached her masterpiece:* Maya Lin, *Smithsonian Magazine,* August 1996.

98 *comments made by a juror:* Maya Lin, "Making the Memorial," *New York Review of Books,* November 2, 2000.

98 *criticized as "too feminine":* Ibid.

98 *called the "space we cannot enter:* Ibid.

98 *After Lin was revealed:* Bill Moyers, "Public Affairs Television 'Becoming American' Interview with Maya Lin," PBS, https://www.pbs.org/becomingamerican/ap_pjourneys_transcript5.html.

102 *Clash!:* Along with Markus's *Clash!,* Richard Nisbett, *The Geography of Thought: How Asians and Westerners Think Differently . . . and Why* (New York: Free Press, 2003), is a terrific plain-language summary of the research on cultural psychology.

102 *engineering advances of the twentieth century:* The American Ceramic Society recognized her invention at its annual meeting in 1931. See also "Anna Wagner Keichline: Bellefonte Architect," Bellefonte Historical and Cultural Association, http://www.bellefontearts.org/local_history_files/local_hist2_new.htm.

102 *made her case to be assigned:* T. Wayne Waters, "Renaissance Woman Anna Keichline," *Town&Gown,* March 2, 2015. To offer her services, Keichline wrote that her architecture background made her suitable for that type of work, but that "if you should deem it advisable to give me something more difficult or as I wish to say more dangerous, I should much prefer it."

102 *Keichline's hollow block required half:* Anna M. Lewis, *Women of Steel and Stone: 22 Inspirational Architects, Engineers, and Landscape Designers* (Chicago: Review Press, 2014).

103 *resulting buildings were more comfortable:* Anna Keichline, "Modern Wall Construction," *Clay Worker,* June 1, 1932.

105 *a key difference in how people approach change:* Nisbett, *The Geography of Thought.*

106 *also signals modesty and shame:* Usha Menon and Richard A. Shweder, "Kali's Tongue: Cultural Psychology and the Power of Shame in Orissa, India," in *Emotion and Culture: Empirical Studies of Mutual Influence* (Washington, D.C.: American Psychological Association, 1994).

107 *logic that led to modern computers:* Chris Dixon, "How Aristotle Created the Computer," *Atlantic,* March 20, 2017.

108 *using treasure they took:* Bruce Johnston, "Colosseum Built with Loot from Sack of Jerusalem Temple," *Telegraph,* June 15, 2001.

109 *across all the major religious texts:* Exodus 22:25, King James Version, broadly defines usury as taking advantage of less fortunate. See also Peter Russell, "Usury—the Root of All Evil?," Spirit of Now, https://www.peterrussell.com/SP/Usury.php. See also Fordham University Center for Medieval Studies, "Medieval Sourcebook: Thomas Aquinas: On Usury, c. 1269–71," and "The Prophet Muhammad's Last Sermon," Internet History Sourcebooks Project, January 2, 2020. See also Bhikkhu Bodhi, "Right Speech, Right Action, Right Livelihood (*Samma Vaca, Samma Kammanta, Samma Ajiva*)," in *The Noble Eightfold Path: The Way to End Suffering* (Onalaska, WA: Buddhist Publication Society Pariyatti Editions, 1994).

109 *the quest for more:* Robert L. Heilbroner, *The Worldly Philosophers: The Lives, Times, and Ideas of the Great Economic Thinkers* (New York: Touchstone, 1999).

Chapter 4: More-ality: Time, Money, and the Modern Gospel of Adding

111 *"Less, but better":* Dieter Rams, *Less but Better* (New York: Gestalten, 2014).

111 *we need to respect simplicity:* George Whitesides, "Toward a Science of Simplicity," TED video, 17:46, https://www.ted.com/talks/george_whitesides_toward_a_science_of_simplicity.

111 *his philosophy on house design:* George Carlin, "George Carlin Talks About 'Stuff,'" YouTube video, 5:10, posted by Tony Gustafsson, May 9, 2012, https://www.youtube.com/watch?v=4x_QkGPCL18.

113 *President Harry Truman's inaugural address:* Harry S. Truman, inaugural address, January 20, 1949, in *Inaugural Addresses of the Presidents of the United States* (Washington, D.C.: Government Printing Office, 1989).

114 *fourth and final point:* Ibid.

115 *no consensus on how to achieve:* Ways to move that altruistic motivation into global policy were increasingly being discussed—including in the July 1944 Bretton Woods Conference, which established the post-WWII international monetary system. But nothing had been decided upon prior to Truman's speech.

116 *look up* usurers *in the Bible:* Ezekiel 18:13, English Standard Version, states, "Lends at interest, and takes profit; shall he then live? He shall not live. He has done all these abominations; he shall surely die; his blood shall be upon himself."

117 *At its depth:* Steve H. Hanke, "A Great Depression," *Globe Asia,* December 2008.

117 *economist John Maynard Keynes:* John Maynard Keynes, *The Essential Keynes*, ed. Robert Skidelsky (New York: Penguin Books, 2015).

118 *infuse government policies:* United States' Employment Act of 1946, P.L. 304, 79th Congress, for example, was committed to "maximum employment, production, and purchasing power."

119 *grew from around $3,000 per year:* "GDP per Capita, 1870 to 2016," Our World in Data, https://ourworldindata.org/economic-growth.

119 *life expectancies have risen:* "Life Expectancy, 1770 to 2015," Our World in Data, https://ourworldindata.org/life-expectancy.

119 *over the age of fifteen could read:* "Literate and Illiterate World Population," Our World in Data, https://ourworldindata.org/literacy.

119 *live in dire poverty:* "World Poverty Clock," World Data Lab, https://worldpoverty.io/.

119 *medieval peasants got more time off:* Lynn Stuart Parramore, "Why a Medieval Peasant Got More Vacation Time Than You," Reuters, August 29, 2013.

121 *Kreider helps readers prove:* Tim Kreider, "The 'Busy' Trap," *New York Times,* June 30, 2012.

121 *needed a stop-doing list:* Jim Collins, *Good to Great: Why Some Companies Make the Leap and Others Don't* (New York: HarperCollins, 2011).

121 *"time famine":* Leslie A. Perlow, "The Time Famine: Toward a Sociology of Work Time," *Administrative Science Quarterly* 44, no. 1 (1999): 57–81.

122 *Perlow first focused on software engineers:* Perlow's engineers were working on software for a budget color laser printer for offices.

122 *glorified their long hours:* Tracy Kidder, *The Soul of a New Machine* (New York: Avon, 1982). See also Fred Moody, *I Sing the Body Electronic: A Year with Microsoft on the Multimedia Frontier* (New York: Viking, 1995). See also G. Pascal Zachary, *Show Stopper!: The Breakneck Race to Create Windows NT and the Next Generation at Microsoft* (New York: Free Press, 1994).

123 *relationships outside of work:* Leslie A. Perlow, "Boundary Control: The Social Ordering of Work and Family Time in a High-Tech Corporation," *Administrative Science Quarterly* 43, no. 2 (1998): 328–57.

123 *quiet time was a stop-doing:* Other tips are in Leslie A. Perlow, *Sleeping with Your Smartphone: How to Break the 24/7 Habit and Change the Way You Work* (Boston, MA: Harvard Business School Publishing Corporation, 2012), and Leslie A. Perlow and Jessica L. Porter, "Making Time Off Predictable—and Required," *Harvard Business Review* 87, no. 10 (2009): 102–9.

123 *army officers have been caught:* Leonard Wong and Stephen J. Gerras, "Lying to Ourselves: Dishonesty in the Army Profession" (Carlisle, PA: Strategic Studies Institute, United States Army War College, February 2015).

123 *fit 297 days of mandatory activities:* Leonard Wong, "Stifled Innovation? Developing Tomorrow's Leaders Today" (Carlisle, PA: Strategic Studies Institute, United States Army War College, April 2012).

124 *The* Code *has ballooned:* "Total Pages in the Code of Federal Regulations and the Federal Register," Regulatory Studies Center, George Washington University, https://regulatorystudies.columbian.gwu.edu/reg-stats.

124 *Clean Air Act:* Clean Air Act (CAA) 42 U.S.C. 7401, enacted in 1970, is a Federal law regulating all sources of air emissions.

125 *"worth crying over spilled milk":* Barack Obama, "Remarks by the President in State of the Union Address," White House, Office of the Press Secretary, January 24, 2012.

125 *each federal agency should:* "Executive Order 13563—Improving Regulation and Regulatory Review," White House, Office of the Press Secretary, January 18, 2011.

126 *prevented harmful industrial pollution:* "Milk Exemption under the SPCC Rule," Environmental Protection Agency (EPA), https://www.epa.gov/oil-spills-prevention-and-preparedness-regulations/milk-exemption-under-spcc-rule.

126 *"relieve a potential burden":* "EPA Updates SPCC Regulation to Exclude Milk and Milk Products / Updated Rule in Keeping with President's Executive Order on Regulatory Reform," EPA Press Office, April 12, 2011.

127 *including in front of Congress:* "Administrator Lisa P. Jackson, Testimony Before the U.S. House Committee on Agriculture," EPA Press Office,

March 3, 2011, in which Jackson stated, "And finally is the notion that EPA intends to treat spilled milk in the same way as spilled oil. This is simply incorrect. Rather, EPA has proposed, and is on the verge of finalizing an exemption for milk and dairy containers. This exemption needed to be finalized because the law passed by Congress was written broadly enough to cover milk containers. It was our work with the dairy industry that prompted EPA to develop an exemption and make sure the standards of the law are met in a commonsense way. All of EPA's actions have been to exempt these containers. And we expect this to become final very shortly."

127 *projected to exceed $1 billion:* "Administrator Lisa P. Jackson, Testimony Before the U.S. House Committee on Agriculture," 2011, stated it would potentially save "the milk and dairy industries more than $140 million a year." For an insider account of the story see also Cass R. Sunstein, *Simpler: The Future of Government* (New York: Simon & Schuster, 2013), which estimated saving a projected $700 million over five years.

128 *comedy often comes from the unexpected:* John Morreall, *The Philosophy of Laughter and Humor* (Albany, NY: SUNY Press, 1986), explains incongruity theory.

129 *shark had not bitten her until:* "Surfer Bitten by Shark: 'I Just Feel Really, Really, Lucky,'" *Hawaii News Now,* March 23, 2006. See also Eloise Aguiar, "North Shore Surfer Survives Shark Bite," *Honolulu Advertiser,* March 24, 2006.

129 *Spending money to save time:* Hal E. Hershfield et al., "People Who Choose Time Over Money Are Happier," *Social Psychological and Personality Science* 7, no. 7 (May 2016).

129 *shockingly small percentage of people:* Ashley V. Whillans et al., "Buying Time Promotes Happiness," *Proceedings of the National Academy of Sciences* (July 2017).

130 *reported greater life satisfaction:* Ibid. asked these questions: "Taking all things together, how happy would you say you are?; 0 = Not at All, 10 = Extremely." And where do you "currently stand in life on a ladder spanning from the worst possible to the best possible life imaginable?; 0 = Bottom Rung, 10 = Top Rung."

132 *subtraction of the pedals:* Drake Baer, "This Dad Built a $10 Million Business by Reinventing the Bicycle," *Business Insider,* May 20, 2014.

133 *largest sportswear company in the world:* "Leading Athletic Apparel, Accessories and Footwear Companies Worldwide in 2020, by Sales (in Million U.S. Dollars)," Statista, https://www.statista.com/statistics/900271/leading-sportswear-and-performance-wear-companies-by-sales-worldwide/.

133 *Marion Rudy had pitched his air concept:* "A Brief History of Nike Air," Nike News, March 18, 2019.

133 *moment in sneaker history:* Tess Reidy, "Nike's Iconic Air Max Trainer Celebrates 25th Anniversary with Tinker Hatfield," *Guardian,* December 14, 2013.

134 *database of the patents issued:* See Google Patents, https://patents.google.com/.

134 *eight closest synonyms:* Katelyn Stenger, Clara Na, and Leidy Klotz, "Less Is More? In U.S. Patents, Design Transformations That Add Occur More Often Than Those That Subtract," Ninth International Conference on Design Computing and Cognition, July 2020. The nine adding terms turned out to be: *add, attach, augment, bolster, coat, connect, join, multi,* and *reinforce.* The nine subtracting terms were: *subtract, detach, free, less, limit, no, remove, simplify,* and *without.*

137 *genesis of his bike epiphany:* "Inventor's Story," Strider, https://www.striderph.com/about.

137 *describing how he subtracted to find:* Ibid.

138 *Truman encouraged his worldwide audience:* Truman, inaugural address.

Chapter 5: Noticeable Less: Finding and Sharing Subtraction

143 *admitted to writing more:* "If I Had More Time, I Would Have Written a Shorter Letter," Quote Investigator, https://quoteinvestigator.com/2012/04/28/shorter-letter/.

143 *"I am now too lazy":* "The Epistle to the Reader," in *The Works of John Locke Esq.,* vol. 1 (London: John Churchill and Sam. Manship, 1714).

144 *Simon named this tendency "satisficing":* Herbert A. Simon, "Rational Choice and the Structure of the Environment," *Psychological Review* 63, no. 2 (1956): 129–38. See also Herbert A. Simon, *Administrative Behavior: A Study of Decision-Making Processes in Administrative Organization* (New York: Macmillan, 1947).

146 *bringing math and science to graphic design:* Joshua Yaffa, "The Information Sage," *Washington Monthly,* May/June 2011.

146 *the term* chartjunk: Edward Tufte, *The Visual Display of Quantitative Information*, 2nd ed. (Cheshire, CT: Graphic Press, 2001).

147 *"Erase non-information-ink"*: Ibid., 105.

147 *"Galileo of Graphics"*: Adam Ashton, "Tufte's Invisible Yet Ubiquitous Influence," *Bloomberg*, June 10, 2009.

147 *"Da Vinci of Data"*: Deborah Shapley, "The Da Vinci of Data," *New York Times*, March 30, 1998.

148 *Lexington, Kentucky, was home:* "Lexington, Kentucky Population 2020," World Population Review, https://worldpopulationreview.com/us-cities/lexington-ky-population.

148 *faster than they could be buried:* Terry Foody, *The Pie Seller, The Drunk, and the Lady: Heroes of the 1833 Cholera Epidemic in Lexington, Kentucky* (n.p.: Terry Foody, 2014), describes how when regular grave diggers fled the city, William Solomon became a local hero for picking up the slack. Solomon drank mostly whiskey, which limited his exposure to cholera. He worked alongside a U.S. army lieutenant named Jefferson Davis, who would go on to be president of the Confederacy. See also "Solomon, William 'King,'" Lexington History Museum, http://lexhistory.org/wikilex/solomon-william-king.

148 *covered the Town Branch Creek:* In post-cholera Lexington, the Town Branch was increasingly viewed as a public health threat. Plus, the city was growing. And, as tended to be the case, when the railroad came to Lexington, it ran alongside the creek, which made adjacent land more valuable. With more activity along the Town Branch, flooding not only spread disease, but it also disrupted commerce and flow of traffic.

149 *The winner was a surprise:* "Town Branch Commons, Lexington, KY," SCAPE, https://www.scapestudio.com/projects/reviving-town-branch/.

149 *soothing beauty of water:* Wallace J. Nichols, *Blue Mind: The Surprising Science That Shows How Being Near, In, On, or Under Water Can Make You Happier, Healthier, More Connected, and Better at What You Do* (Boston: Little, Brown, 2014).

149 *Minetta Brook remains hidden:* Larissa Zimberoff, "Minetta Brook: A Lost River Under the Streets of Manhattan," Untapped New York, July 24, 2012.

149 *Islais Creek flows under:* Joel Pomerantz, Seep City, http://www.seepcity.org.

152 *practice of taking out parts:* Paul Smith, "Hemingway's Early Manuscripts: The Theory and Practice of Omission," *Journal of Modern Literature* 10, no. 2 (1983): 268–88.

152 *"would strengthen the story":* Ernest Hemingway, *A Moveable Feast* (New York: Scribner's, 1964). Hemingway explains that "I omitted the real end [of 'Out of Season'] which was that the old man hanged himself. This was omitted on my new theory that . . ."

152 *what the research data reveals:* Daniel Oppenheimer, "Consequences of Erudite Vernacular Utilized Irrespective of Necessity: Problems with Using Long Words Needlessly," *Applied Cognitive Psychology* 20 (2006): 139–56.

152 *writing guide:* William Strunk Jr. and E. B. White, *The Elements of Style,* 4th ed. (London: Pearson, 1999).

152 *appears on more course syllabi:* "Most Frequently Assigned Titles," Open Syllabus, https://opensyllabus.org/, showing *The Elements of Style* is the most frequently assigned text in U.S. academic syllabi (as of September 2020).

152 *"Omit needless words":* Strunk Jr. and White, *The Elements of Style.*

153 *our participants tended to add:* When editing another person's summary, just 28 percent of participants subtracted; when editing their own summaries, 14 percent of participants subtracted. For details, see Adams et al., "People Systematically Overlook Subtractive Changes."

153 *an argument's length is often:* Caren M. Rotello and Evan Heit, "Modeling the Effects of Argument Length and Validity on Inductive and Deductive Reasoning," *Journal of Experimental Psychology: Learning, Memory, and Cognition* 35, no. 5 (September 2009): 1317–30.

153 *One of the questions:* "Admissions: Apply," Harvard University Graduate School of Design, https://www.gsd.harvard.edu/admissions/apply/ (as of September 2020).

155 *Darkness on the Edge of Town:* Bruce Springsteen, *Born to Run* (New York: Simon & Schuster, 2016).

155 *describes in his autobiography:* Ibid., 264.

156 *yet to have a top-20 single:* Manfred Mann's Earth Band did take a cover of Springsteen's "Blinded by the Light" to #1.

156 *"nothing less than a breakthrough":* Dave Marsh, "Darkness on the Edge of Town," *Rolling Stone,* July 27, 1978. The reviewer, Dave Marsh, went on to write several biographies about Springsteen.

156 *album of the year:* "Albums and Tracks of the Year: 1978," *NME,* October 10, 2016.

157 *concert set lists are likely:* "Bruce Springsteen Tour Statistics: Songs Played Total," Setlist.fm, https://www.setlist.fm/stats/bruce-springsteen-2bd6dcce.html.

157 *Genius Awards for exceptional creativity:* Adrian Higgins, "For the First Time, MacArthur Foundation has Given 'Genius' Award to a Landscape Architect," *Washington Post,* October 18, 2017.

158 *attached to the work:* Christopher Y. Olivola, "The Interpersonal Sunk-Cost Effect," *Psychological Science* 29, no. 7 (2018): 1072–83.

158 *even less likely to subtract words:* When it was their own writing, only 14 percent shortened the summary to improve it. When it was someone else's, 28 percent shortened. For details, see Adams et al., "People Systematically Overlook Subtractive Changes."

159 *the tidying mogul is famous:* Marie Kondo, *The Life-Changing Magic of Tidying Up* (Berkeley, CA: Ten Speed Press, 2014).

160 *post-book Netflix series: Tidying Up with Marie Kondo* (Los Gatos, CA: Netflix, 2019).

161 *title of her next book:* Marie Kondo, *Spark Joy: An Illustrated Master Class on the Art of Organizing and Tidying Up* (Berkeley, CA: Ten Speed Press, 2016).

162 *self-efficacy is:* Albert Bandura, *Self-Efficacy: The Exercise of Control* (New York: W. H. Freeman, 1997). See also "Information on Self-Efficacy, A Community of Scholars," University of Kentucky, https://www.uky.edu/~eushe2/Pajares/self-efficacy.html. See also James E. Maddux, "Self-Efficacy: The Power of Believing You Can" in *The Oxford Handbook of Positive Psychology,* 2nd ed. (Oxford Handbooks Online, September 2012).

163 *researchers call "flow" states:* Mihaly Csikszentmihalyi, *Finding Flow: The Psychology of Engagement with Everyday Life* (New York: Basic Books, 1997).

164 *the last job he had:* Hardeep Phull, "Springsteen Hasn't Done a Day's Work Since His Teens," *New York Post,* April 29, 2017.

165 *to share his data-ink ratio:* Mark Zachary and Charlotte Thralls, "Cross-Disciplinary Exchanges," *Technical Communication Quarterly* 13, no. 4 (2004): 447–62.

165 *Lewin defined psychological valence:* Kurt Lewin, *Field Theory in Social Science: Selected Theoretical Papers,* ed. Dorwin Cartwright (New York:

Harpers, 1951). See also Kurt Lewin, "Need, Force and Valence in Psychological Fields," in *Classic Contributions to Social Psychology* (London: Oxford University Press, 1972).

165 *classify thousands of words:* Saif M. Mohammad and Peter D. Turney, "Crowdsourcing a Word-Emotion Association Lexicon," *Computational Intelligence* 29, no. 3 (2013): 436–65. See also Saif M. Mohammad and Peter D. Turney, "Emotions Evoked by Common Words and Phrases: Using Mechanical Turk to Create an Emotion Lexicon," NAACL-HLT Conference, January 2010.

166 *four verbs:* Kate Orff et al., "The Deep Section: Karst Urbanism in Town Branch Commons," *Oz* 37, no. 9 (2015). They "aim to reveal the underground stream and the multi-faceted qualities of water as a series of urban destinations, clean Town Branch into an ecologically viable and safe waterway, carve Town Branch into the downtown fabric to stage its topographic qualities, and connect Town Branch back into the neighborhoods where it originates."

167 *value things we have even more:* Amos Tversky and Daniel Kahneman, "Prospect Theory: An Analysis of Decision Under Risk," *Econometrica* 47 (1979): 263–91.

168 *Kahneman's own Nobel Prize in 2002:* Amos Tversky passed away in 1996.

168 *his bestseller:* Daniel Kahneman, *Thinking, Fast and Slow* (New York: Farrar, Straus and Giroux, 2011).

168 *subtractively titled account:* Michael Lewis, *The Undoing Project: A Friendship That Change Our Minds* (New York: W. W. Norton, 2016).

168 *brilliantly simple experiments:* Daniel Kahneman et al., "Experimental Tests of the Endowment Effect and the Coase Theorem," *Journal of Political Economy* 98, no. 6 (December 1990): 1325–48.

168 *must explain the difference:* Couldn't this difference still be chalked up to some of Herbert Simon's findings? Perhaps students just don't want to bother getting out of their seat to trade. Maybe they are shy and don't want to talk to each other. Kahneman and his team ruled out this explanation by running the same experiment, except that they replaced the mugs with tokens, which the students could cash in before leaving class. The tokens require the same moving and talking to trade as the mugs do, but for the tokens, which have no value beyond the classroom, the willingness to pay and the willingness to sell were the same.

168 *Similar loss-averse behavior:* Alan Silberberg et al., "On Loss Aversion in Capuchin Monkeys," *Journal of the Experimental Analysis of Behavior* 89, no. 2 (March 2008): 145–55.

168 *stimulate different circuits in our brains:* Sabrina M. Tom et al., "The Neural Basis of Loss Aversion in Decision-Making Under Risk," *Science* 315, no. 5811 (January 2007): 515–18. See also Ben Seymour et al., "Differential Encoding of Losses and Gains in the Human Striatum," *Journal of Neuroscience* 27, no. 18 (May 2007): 4826–31.

170 *Orff's book:* Kate Orff, *Toward an Urban Ecology* (New York: Monacelli Press, 2016).

Chapter 6: Scaling Subtraction: Using Less to Change the System

171 *"Sun City":* Dave Marsh and James Bernard, *New Book of Rock Lists* (New York: Fireside, 1994).

173 *Sharpeville massacre:* "The Sharpeville Massacre, 1960," Divestment for Humanity: The Anti-Apartheid Movement at the University of Michigan, https://michiganintheworld.history.lsa.umich.edu/antiapartheid/.

173 *recommended cutting off military aid:* United Nations Security Council Resolution 181 (1963) called upon states to voluntarily cease the sale and shipment of all ammunition and military vehicles to South Africa (this clause finally became mandatory with Resolution 418 (1977)).

173 *Ibram Kendi:* Ibram X. Kendi, *How to Be an Antiracist* (New York: One World, 2019).

176 *could use a Galilean revolution:* Kurt Koffka, *Principles of Gestalt Psychology* (London: Routledge, 1935).

176 *represent human behavior as a system:* Kurt Lewin, *Field Theory in Social Science: Selected Theoretical Papers*, ed. Dorwin Cartwright (New York: Harpers, 1951).

177 *his interest in social problems:* Lewin also helped found the Society for the Psychological Study of Social Issues (SPSSI), which stills exists, with aims to encourage research upon those psychological problems most vitally related to modern social, economic, and political policies, and to help the public and its representatives understand and use contributions from the scientific investigation of human behavior in the formation of social policies.

177 *changing the invisible forces:* "Making the Invisible Visible: Transformative Research and Social Action," SPSSI 2020 Summer Conference, Denver, Colorado, June 26–28, 2020.

178 *Kahneman put it this way:* Stephen Dubner, "How to Launch a Behavior-Change Revolution," *Freakonomics Radio,* episode 306, produced by Stephen J. Dubner, 47:53.

180 *repeatedly clarified, to no avail:* Russell A. Dewey, "Gestalt Psychology," in *Psychology: An Introduction,* https://www.psywww.com/intropsych/.

180 *Braess's math plays out:* Dietrich Braess, "On a Paradox of Traffic Planning," *Transportation Science* 39, no. 4 (November 2005): 443–556.

181 *not limited to roads and traffic:* Richard Steinberg and Willard I. Zangwill, "The Prevalence of Braess' Paradox," *Transportation Science* 17, no. 3 (August 1983): 239–360.

181 *found in electrical power grids:* Joel E. Cohen and Paul Horowitz, "Paradoxical Behaviour of Mechanical and Electrical Networks," *Nature* 352 (1991): 699–701.

181 *biological systems:* Sagra Sahasrabudhe and Adilson E. Motter, "Rescuing Ecosystems from Extinction Cascades Through Compensatory Perturbations," *Nature Communications* 2, no. 170 (January 2011): 1–8.

182 *we won the tournament:* "2019 Patriot League Men's Soccer Record Book," Patriot League, https://patriotleague.org/sports/2016/6/13/sports-m-soccer-archive-patr-m-soccer-archive-html.aspx.

183 *A typical project I worked on:* "Elizabeth: Dr. Albert Einstein School (aka #29)," State of New Jersey Schools Development Authority, https://www.njsda.gov/NJSDA/ProjectSchoolDetails/SchoolGrantDetails?vProjectID=39–1320-x05&vSchoolDistrict=Elizabeth.

185 *much larger school construction program:* State of New Jersey Schools Development Authority, https://www.njsda.gov.

187 *quest to make a perfect map:* Jorge Luis Borges, purportedly from *Suárez Miranda, Travels of Prudent Men, Book Four, Ch. XLV* (Lérida: 1658) in Jorge Luis Borges, "On Exactitude in Science," in *Collected Fictions*, trans. Andrew Hurley (n.p.: Penguin Books: 1999). See also Lewis Carroll, *Sylvie and Bruno Concluded, Chapter XI* (London: 1895).

187 *ends up blocking out the sun:* Borges, *Collected Fictions,* stated, "In that Empire, the Art of Cartography attained such Perfection that the map of a single Province occupied the entirety of a City, and the map of the Empire, the entirety of a Province. In time, those Unconscionable

Maps no longer satisfied, and the Cartographers Guilds struck a Map of the Empire whose size was that of the Empire, and which coincided point for point with it. The following Generations, who were not so fond of the Study of Cartography as their Forebears had been, saw that that vast Map was Useless, and not without some Pitilessness was it, that they delivered it up to the Inclemencies of Sun and Winters. In the Deserts of the West, still today, there are Tattered Ruins of that Map, inhabited by Animals and Beggars; in all the Land there is no other Relic of the Disciplines of Geography."

187 *working memory hits limits:* George A. Miller, "The Magical Number Seven, Plus or Minus Two: Some Limits on Our Capacity for Processing Information," *Psychological Review* 63, no. 2 (March 1956): 81–97.

188 *as he hoped it would:* Ibid., 96, stated, "In fact, I feel that my story here must stop just as it begins to get really interesting."

188 *severely restricted capacity, often* below *seven:* See Alan Baddeley, "The Magical Number Seven: Still Magic After All These Years?," *Psychological Review* 101, no. 2 (1994): 353–56. See also Paul M. Bays and Masud Husain, "Dynamic Shifts of Limited Working Memory Resources in Human Vision," *Science* 321, no. 5890 (August 2008): 851–54. See also Wei Ji Ma et al., "Changing Concepts of Working Memory," *Nature Neuroscience* 17 (2014): 347–56.

189 *Her timeless book:* Donella H. Meadows, *Thinking in Systems: A Primer,* ed. Diana Wright (White River Junction, VT: Chelsea Green, 2008).

191 *guides them through some version:* Agency for Healthcare Research and Quality (AHRQ), *Emergency Severity Index (ESI): A Triage Tool for Emergency Departments,* version 4 (Rockville, MD: AHRQ, last reviewed May 2020).

192 *subtracting detail could save lives:* Peter Pronovost's central catheter case, and others like it, are described in Atul Gawande, *The Checklist Manifesto: How to Get Things Right* (New York: Picador, 2011). Gawande's book has lived up to its title, sparking a revolution in public health. And the same subtracting that brought lifesaving checklists can also enhance our thinking about less. The doctor and author put it well: "Under conditions of true complexity—where the knowledge required exceeds that of any individual and unpredictability reigns—efforts to dictate every step from the center will fail. People need room to act and adapt."

192 *causing about thirty thousand deaths each year:* According to Peter Pronovost et al., "An Intervention to Decrease Catheter-Related Bloodstream Infections in the ICU," *New England Journal of Medicine* 355 (December 2006): 2725–32, there are "up to 28,000 deaths among patients." Compare that to *Traffic Safety Facts: 2006 Data* (Washington, D.C.: NHTSA National Center for Statistics and Analysis, 2006), which reported 32,119 motor vehicle fatalities in 2006 (this number does not include motorcyclists or non-occupants).

192 *insert a catheter line:* "Appendix 5: Central Line Insertion Care Team Checklist," AHRQ, https://www.ahrq.gov/hai/clabsi-tools/appendix-5.html.

192 summary *of the guidelines:* "Practice Guidelines for Central Venous Access: A Report by the American Society of Anesthesiologists Task Force on Central Venous Access," *Anesthesiology* 116, no. 3 (March 2012): 539–73.

193 *Michigan and Rhode Island:* Pronovost et al., "An Intervention to Decrease."

193 *surgical fires each year:* "Surgical Fire Prevention," ECRI, https://www.ecri.org/solutions/accident-investigation-services/surgical-fire-prevention.

193 *earned all As in college:* To the great delight of my brother and me, my sister earned a single A in her final semester.

194 *Jenga originated:* Leslie Scott, *About Jenga: The Remarkable Business of Creating a Game that Became a Household Name* (Austin, TX: Greenleaf Book Group Press, 2010).

195 *desire to build:* The name Jenga is derived from *kujenga*, a Swahili word meaning "to build."

195 *lead to entirely different outcomes:* This sequential processing is so neglectful of subtraction that it is called *preference construction.* See Paul Slovic, "The Construction of Preference," *American Psychologist* 50, no. 5 (1995): 364–71.

195 *Project management textbooks:* Chris Hendrickson, *Project Management for Construction: Fundamental Concepts for Owners, Engineers, Architects and Builders,* version 2.2 (n.p., 2008), https://www.profkrishna.com/ProfK-Assets/HendricksonBook.pdf.

196 *story dates the innovation to 1847:* The American Donut Corp sponsored the Great Donut Debate at the Astor Hotel in New York (about five miles from where the Collyer brothers lived) with the subject "Who

put the hole in the doughnut?" Gregory's cousin brought letters and signed affidavits, the celebrity judges declared him the winner, and so we'll use that story here. See also Candy Sagon, "The Hole Story," *Washington Post,* March 6, 2002.

197 *plenty of appeal on their own:* This assumes you are using a ring-cut process, which is typical with the cakey variety. When doughnuts are extruded, as in Levitt's process, there is not actually a hole, but the dough does go further. See David A. Taylor, "The History of the Doughnut," *Smithsonian Magazine,* March 1998.

Chapter 7: A Legacy of Less: Subtracting in the Anthropocene

199 *dominant influence on our planet's well-being:* Simon L. Lewis and Mark A. Maslin, "Defining the Anthropocene," *Nature* 519 (2015): 171–80.

199 *here's what you need to know:* Dr. Seuss, *The Lorax* (New York: Random House, 1971). Seuss's classic has since sold more than two million copies and, in 2012, was produced by Universal Pictures, Illumination Entertainment, and Dr. Seuss Enterprises as an animated feature film.

200 *innovative members of his species:* Ray Anderson, *Mid-Course Correction: Toward a Sustainable Enterprise: The Interface Model* (n.p.: Peregrinzilla Press, 1999).

201 *atmospheric concentration of carbon dioxide:* "Climate Change: How Do We Know?," NASA, http://climate.nasa.gov/evidence. See also "Is the Current Level of Atmospheric CO_2 Concentration Unprecedented in Earth's History?," National Academies Press, https://www.nap.edu/resource/25733/interactive/.

201 *begun to harness the power:* "History of Energy Consumption in the United States, 1775–2009," U.S. Energy Information Administration, https://www.eia.gov/todayinenergy/detail.php?id=10.

201 *parts per million—and rising:* "CO_2 Ice Core Data," CO_2.earth, https://www.co2.earth/co2-ice-core-data.

202 *250,000 deaths every year:* "Climate Change and Health," World Health Organization, February 1, 2018, https://www.who.int/news-room/fact-sheets/detail/climate-change-and-health.

202 *a conservative estimate:* Andy Haines and Kristie Ebi, "The Imperative for Climate Action to Protect Health," *New England Journal of Medicine* 380 (January 2019): 263–73.

202 *whose image reminded me:* "Alan Kurdi," 100 Photographs, *Time,* 2005, http://100photos.time.com/photos/nilufer-demir-alan-kurdi.

203 *drought made worse by climate change:* Mark Fischetti, "Climate Change Hastened Syria's Civil War," *Scientific American,* March 2, 2015.

204 *one of the most influential:* Jørgen Stig Nørgård et al., "The History of the Limits to Growth," *Solutions* 1, no. 2 (March 2010): 59–63.

204 *researchers' conclusions to their essence:* Donella H. Meadows et al., *The Limits to Growth* (New York: Report to the Club of Rome, 1972). Despite the ominous title, one of the three scenarios in *Limits* is an optimistic sketch of how humanity could continue thriving without pillaging the planet.

204 *"the greatest good":* Gifford Pinchot, *Breaking New Ground* (Washington, D.C.: Island Press, 1998), states, "Conservation is the foresighted utilization, preservation and/or renewal of forests, waters, lands and minerals for the greatest good of the greatest number for the longest time."

205 *exceeding the safe operating conditions:* Will Steffen et al., "Planetary Boundaries: Guiding Human Development on a Changing Planet," *Science* 347, no. 6223 (February 2015): 1259855.

205 *limit economic growth:* For example, Nicolas Stern, *The Economics of Climate Change: The Stern Review* (Cambridge, UK: Cambridge University Press, 2007).

205 *distrust in climate science*: Justin Farrell, "Corporate funding and ideological polarization about climate change," *Proceedings of the National Academies of Sciences* 113, no. 1 (2016): 92–7.

206 *Tufte-esque graphics of global progress:* Hans Rosling et al., *Factfulness: Ten Reasons We're Wrong About the World—and Why Things Are Better Than You Think* (New York: Flatiron Books, 2018).

206 *far on one side or the other:* The wizard-versus-prophet argument is a theme of John McPhee, *Encounters with the Archdruid* (New York: Farrar, Straus and Giroux, 1971). See also, Charles Mann, *The Wizard and Prophet: Two Remarkable Scientists and Their Dueling Visions to Shape Tomorrow's World* (New York: Knopf, 2018).

208 *disproportionately on economically poor minorities:* S. Nazrul Islam and John Winkel, "Climate Change and Social Inequality," working paper no. 152 (New York: United Nations, October 2017).

208 *One of the IPCC's latest reports:* IPCC, *Climate Change 2014: Synthesis Report. Contribution of Working Groups I, II, and III to the Fifth Assessment*

Report of the Intergovernmental Panel on Climate Change (Geneva, Switzerland: IPCC, 2015).

208 *countless interdependent issues:* And then some goals and situations are linked closely to those for climate. The extinction equivalent of the IPCC is the Intergovernmental Panel on Biodiversity and Ecosystem Services (IPBES), which has an equally rigorous process and equally dense report.

209 *goals had to also consider impacts:* Carolina Mauri, "Co$_2$sta Rica, The National Climate Change Strategy and the Carbon Neutrality Challenge," Ministry of Environment, Energy, and Telecommunications, https://unfccc.int/files/meetings/sb30/press/application/pdf/session4_mitigation.pdf.

210 *first carbon-neutral country:* John McPaul, "Costa Rica Pledges to Be 'Carbon Neutral' by 2021," Reuters, June 7, 2007.

210 *biofuels:* Biofuels are consumed more slowly than regeneration rate.

210 *doctrine "Reduce, reuse, recycle" arose:* "'Reduce, Reuse, Recycle' Button," Smithsonian Natural Museum of American History, https://americanhistory.si.edu/collections/search/object/nmah_1284430.

211 *restoring forests:* Jean-Francois Bastin et al., "The Global Tree Restoration Potential," *Science* 365, no. 6448 (July 2019): 76–79. See also "Erratum for the Report," *Science* 368 no. 6494 (May 2020): eabc8905.

212 *spray aerosols into the atmosphere:* Justin McClellan et al., "Cost Analysis of Stratospheric Albedo Modification Delivery Systems," *Environmental Research Letters* 7, no. 3 (August 2012): 034019.

212 *lighten the color:* Hashem Akbari and H. Damon Matthews, "Global Cooling Updates: Reflective Roofs and Pavements," *Energy and Buildings* 55 (December 2012): 2–6.

212 *dump tons of iron filings:* Mark G. Lawrence et al., "Evaluating Climate Geoengineering Proposals in the Context of the Paris Agreement Temperature Goals," *Nature Communications* 9, no. 3734 (2018).

213 *some plans do put subtracting first:* Engineers are working on giant air-filtering vacuums, artificial trees, and new types of concrete and crops that could extract and store carbon dioxide.

214 *save more money:* Hal E. Hershfield et al., "Increasing Saving Behavior Through Age-Progressed Renderings of the Future Self," *Journal of Marketing Research* 48 (2019): S23–S37.

214 *the practice of "visioning":* Arnim Wiek and David Iwaniec, "Quality Criteria for Visions and Visioning in Sustainability Science," *Sustainability Science* 9, no. 4 (2014): 497–512.

214 *Costa Rica reducing emissions:* Hannah Ritchie and Max Roser, "Costa Rica: What Share of Global CO_2 Emissions Are Emitted by the Country?," Our World in Data, https://ourworldindata.org/co2/country/costa-rica, shows that in 2006, Costa Rica was 0.02 percent, rising to 0.03 percent in 2007 and following an overall downward trajectory since.

215 *they won't change the Anthropocene:* Hannah Ritchie and Max Roser, "CO_2 Emissions," Our World in Data, https://ourworldindata.org/co2-emissions.

215 *carbon neutrality by 2050:* Megan Darby and Isabelle Gerretsen, "Which Countries Have a Net Zero Carbon Goal?," *Climate Change News,* https://www.climatechangenews.com/2020/09/17/countries-net-zero-climate-goal/.

215 *progress on the path to less:* "Country Summary: Costa Rica," Climate Action Tracker, https://climateactiontracker.org/countries/costa-rica/, states that the country generates 98 percent of its electricity from renewable sources.

215 *earned Costa Rica recognition:* Anna Bruce-Lockhart, "Which Is the Greenest, Happiest Country in the World," World Economic Forum, July 29, 2016. See also "Costa Rica," Happy Planet Index, http://happyplanetindex.org/countries/costa-rica.

216 *thousands of times higher:* Gerardo Ceballos et al., "Accelerated Modern Human-Induced Species Losses: Entering the Sixth Mass Extinction," *Science Advances* 1, no. 5 (Jun 2015): e1400253.

217 *GDP hasn't always been:* Elizabeth Dickinson, "GDP: A Brief History," *Foreign Policy,* January 3, 2011.

218 *GDP struggles to measure:* "The Trouble with GDP," *Economist,* April 30, 2016.

219 *measure what matters:* John Maynard Keynes, one of the men we have to thank for GDP, had some wise advice for those faced with things that were hard to measure: "I would rather be vaguely right, than precisely wrong."

219 *Using this modified measure:* Costa Rica has been reinvesting in less for a long time. In 1997, they established the first national-scale program that paid for mitigation of greenhouse gas emissions. Other nations

have conservation subsidies, but Costa Rica was the first to reframe them as payments for services. It wasn't perfect. The flat-rate payment scheme doesn't consider the degree of environmental service provided, which means some of the highest-value areas remain outside of the program. Plus, the fuel taxes from which the incentives were provided had to be supplemented by outside funding from the World Bank and others. Plus, it only helps landowners. See Katia Karousakis, "Incentives to Reduce GHG Emissions from Deforestation: Lessons Learned from Costa Rica and Mexico," Organisation for Economic Co-Operation and Development and International Energy Agency, May 2007.

219 *got rid of its army:* Judith Eve Lipton and David P. Barash, *Strength Through Peace: How Demilitarization Led to Peace and Happiness in Costa Rica, and What the Rest of the World can Learn From a Tiny, Tropical Nation* (Oxford, UK: Oxford University Press, 2019).

219 *Its GDP is around:* "GDP Per Capita (Current US$)—Costa Rica, United States," World Bank, https://data.worldbank.org/indicator/NY.GDP.PCAP.CD?locations=CR-US.

219 *longer lives, on average, than Americans:* Luis Rosero-Bixby and William H. Dow, "Exploring Why Costa Rica Outperforms the United States in Life Expectancy: A Tale of Two Inequality Gradients," *Proceedings of the National Academy of Sciences* 113, no. 5 (Feb 2016): 1130–37.

220 *offered a general treatment:* Results of the Resilience Project are summarized in Lance Gunderson and Crawford Holling, eds., *Panarchy: Understanding Transformations in Human and Natural Systems* (Washington, D.C.: Island Press, 2009), and the quote is from a summary of the book and project by Crawford Holling, "Understanding the Complexity of Economic, Ecological, and Social Systems," *Ecosystems* 4, no. 5 (2001): 390–405.

221 *vocal and influential advocate:* Desmond Tutu, "We Need An Apartheid-Style Boycott to Save the Planet," *Guardian,* April 10, 2014.

221 *five times as many emissions:* "Do the Math," 350.org, https://math.350.org/.

222 *divest their holdings in fossil fuel:* Teresa Watanabe, "UC Becomes Nation's Largest University to Divest Fully from Fossil Fuels," *Los Angeles Times,* May 19, 2020.

222 *pope has called on all Catholics:* Philip Pullella, "Vatican Urges Catholics to Drop Investments in Fossil Fuels, Arms," Reuters, June 18, 2020.

222 *among those that have:* Track additional progress at 350.org, https://www.350.org.

223 *Truman acknowledged this point:* Harry S. Truman, inaugural address, January 20, 1949, in *Inaugural Addresses of the Presidents of the United States* (Washington, D.C.: Government Printing Office, 1989).

Chapter 8: From Information to Wisdom: Learning by Subtracting

224 *downsides of being the latest generation:* Cal Newport, "Is Email Making Professors Stupid?," *Chronicles of Higher Education,* February 12, 2019.

225 *one hundred thousand words a day:* Nick Bilton, "The American Diet: 43 Gigabytes a Day," *New York Times,* December 9, 2009.

225 *one internet minute:* Jeff Desjardin, "What Happens in an Internet Minute in 2017?," *Visual Capitalist,* August 2, 2017.

225 *information threatens our mental health:* Barry Schwartz, *The Paradox of Choice: Why More Is Less* (New York: HarperCollins, 2004).

226 *participation required for a functioning democracy:* Cass R. Sunstein, *How Change Happens* (Cambridge, MA: MIT Press, 2019).

226 *one face-to-face conversation at a time:* Daniel J. Levitin, "Why It's So Hard to Pay Attention, Explained by Science," *Fast Company,* August 23, 2015.

226 *"a wealth of information creates":* Herbert A. Simon, "Designing Organizations for an Information-Rich World," in *Computers, Communications, and the Public Interest* (Baltimore, MD: Johns Hopkins University Press, 1971).

226 *relationship between economic poverty and bad decisions:* Sendhil Mullainathan and Eldar Shafir, *Scarcity: Why Having Too Little Means So Much* (New York: Picador, 2013).

227 *growth of information outpaces:* Richard Van Noorden, "Global Scientific Output Doubles Every Nine Years," *Nature News Blog,* May 7, 2014. See also EIA, *International Energy Outlook 2019 With Projections to 2050* (Washington, D.C.: U.S. Energy Information Administration, 2019). See also "World Economic Outlook Reports," International Monetary Fund, https://www.imf.org/en/Publications/WEO.

227 *subtracting less useful information:* John M. Keynes, *A Treatise on Probability* (London: Macmillan, 1921).

228 *was already warning:* Ecclesiastes 12:12, Christian Standard Bible.

228 *end-of-life moral advice letters:* Lucius Annaeus Seneca, *Letters from a Stoic,* trans. Robin Campbell (New York: Penguin Books, 1969).

228 *ways to store, summarize, and sort:* Ann M. Blair, *Too Much to Know: Managing Scholarly Information Before the Modern Age* (New Haven, CT: Yale University Press, 2010).

229 *simplest selection filter:* Richard Saul Wurman, *Information Anxiety* (New York: Bantam, 1990).

230 *cheating:* Clara Levy, "Cheating at Mines: Part One & Two," *Oredigger,* March 8, 2016, http://oredigger.net/2016/03/cheating-at-mines-part-one/.

232 *process brought librarians to tears:* Helen Carter, "Authors and Poets Call Halt to Book Pulping at Manchester Central Library," *Guardian,* June 22, 2012.

236 *"big-kids" preschool:* St. Mark Lutheran Preschool, http://www.stmarkpreschool.net/.

237 constructivism: "Workshop: Constructivism as a Paradigm for Teaching and Learning," WNET Education, https://www.thirteen.org/edonline/concept2class/constructivism/, provides a good overview which began with Piaget. This is not to be confused with Papert's even more adding-centric "constructionism" explained in Idit Harel and Seymour Papert, *Constructionism* (Norwood, NJ: Ablex Publishing, 1991). "From constructivist theories of psychology we take a view of learning as a reconstruction rather than as a transmission of knowledge. Then we extend the idea of manipulative materials to the idea that learning is most effective when part of an activity the learner experiences as constructing a meaningful product." It then adds the idea that this happens best when the learner is "consciously engaged in constructing a public entity, whether it's a sand castle on the beach or a theory of the universe."

238 *new ideas stand on old ones:* Dedre Gentner and Albert L. Stevens, eds., *Mental Models* (New York: Psychology Press, 2014).

239 *gives us our unique human advantages:* Andrea A. DiSessa, "A History of Conceptual Change Research: Threads and Fault Lines" (Berkeley: UC–Berkeley, 2014).

239 *distort how we learn:* For one example, see Michael Allen, *Misconceptions in Primary Science* (New York: McGraw-Hill, 2010).

240 *joined a cult:* Leon Festinger et al., *When Prophecy Fails: A Social and Psychological Study of a Modern Group That Predicted the Destruction of the World* (Minneapolis: University of Minnesota Press, 1956).

241 *we bend both instead:* John D. Sterman and Linda Booth Sweeney, "Understanding Public Complacency About Climate Change: Adults' Mental Models of Climate Change Violate Conservation of Matter," *Climatic Change* 80 (2007): 213–38.

241 *put forth an alternative approach:* John P. Smith III et al., "Misconceptions Reconceived: A Constructivist Analysis of Knowledge in Transition," *The Journal of Learning Sciences* 3, no. 2 (1994): 115–63.

242 *process of reframing our ideas:* George J. Posner et al., "Accommodation of a Scientific Conception: Toward a Theory of Conceptual Change," *Science Education* 77, no. 2 (April 1982): 211–27.

242 *how we construct new knowledge:* As the researchers put it: "When we act on the expectation that the world operates in one way and it violates our expectations, we often fail, but by accommodating this new experience and reframing our model of the way the world works, we learn from the experience of failure."

243 *"revolutionary" progress:* Thomas S. Kuhn, *The Structure of Scientific Revolutions,* 3rd ed. (Chicago: University of Chicago Press, 1996).

243 *most influential of the last century:* "100 Best Nonfiction," Modern Library, https://www.modernlibrary.com/top-100/100-best-nonfiction/.

244 *expertise in cognitive science:* Nancy J. Nersessian, *Creating Scientific Concepts* (Cambridge, MA: MIT Press, 2010).

245 *when we learn by analogy:* Dedre Gentner and Keith J. Holyoak, "Reasoning and Learning by Analogy: Introduction," *American Psychologist* 52, no. 1 (1997): 32–34.

245 *new evidence fails to remove misconceptions:* For one example see Dedre Gentner, "Flowing Waters or Teeming Crowds: Mental Models of Electricity," in *Mental Models* (Hillsdale, NJ: Lawrence Erlbaum Associates, 1983), 99–129. See also David E. Brown and John Clement, "Overcoming Misconceptions via Analogical Reasoning: Abstract Transfer Versus Explanatory Model Construction," *Instructional Science* 18 (1989): 237–61. See also David E. Brown, "Using Examples and Analogies to Remediate Misconceptions in Physics: Factors Influencing Conceptual Change," *Journal of Research in Science Teaching* 29, no. 1 (1992): 17–34.

About the Author

Leidy Klotz studies how we transform things from how they are to how we want them to be. His research on the science of design has appeared in both *Nature* and *Science*, and he has written for *The Washington Post*, *Fast Company*, *Lit Hub*, and *Behavioral Scientist*. A professor appointed in engineering, architecture, and business at the University of Virginia, Leidy has authored more than eight original research articles and secured more than $10 million in competitive funding to support his and others' work in this area. Before becoming a professor, Leidy played professional soccer.